INEXPENSIVE CNC PROJECTS

Robert J Davis II

Inexpensive CNC Projects
Copyright 2014-2016 Robert J Davis II
All rights reserved
Copying by permission only, except for short quotes

Update

This is an updated version with an improved metal design, with an Arduino based USB interface, and laser engraving support.

Disclaimer

The construction and safe operation of these devices is the sole responsibility of the reader. If you do not know what you are doing then do not try to build these projects. You could start with the simpler projects found in my books "Arduino LED Projects" or "Arduino Robotics Projects".

This book is intended to help you to get started building your own CNC machine. Feel free to improve upon my designs to make a better machine. It is not meant to be a professional machine and at this point can only work on wood and plastic products. By upgrading it with metal parts a more powerful CNC machine can be built.

Dedication

I would like to dedicate this book to my dad, Robert J Davis Sr. He was a fantastic CNC programmer who did most of the CNC programming in his head.

I remember dad decorating the house for Christmas with red and green Mylar "paper tape" that once held computer programs that he brought home from work. I remember dad trying to explain to me how "right shift and add" was multiplying and "left shift and subtract" is dividing to a computer.

My dad also taught me everything from how to do electrical wiring, to home construction, to how to make many car repairs. I remember dad test driving a car and then it broke down out in the middle of nowhere. We walked to a farmhouse singing "Ford, Ford it is the best ride a mile and walk the rest". I have many fond memories of my dad. He passed away shortly after this book was first published.

Table of Contents

Background.. 4

1. Design Concepts.. 6

2. Materials used (Bill of Materials)...................................... 22
    Needed Parts - Optional Parts - Wasted Money

3. Mechanical design... 33
    Y Axis Parts - X Axis Parts - Z Axis Parts
    Electronics Board

4. Mechanical assembly... 45
    Base or Y axis - Gantry or X axis - Z axis

5. The Electronics... 53
    L297 + TIP120 - A4988 "Step Stick" - TB6560
    BOB (Break out Board) - Home/Limit Switches

6. CNC Pendant... 67
    X, Y, and Z switches - 555 Automatic Clock

7. Linux CNC Program... 70
    Basic Machine Setup – Parallel Port – X, Y, and Z axis

8. CNC Commands... 80
    A-Z Commands - G codes - M codes
    CNC editor - Sample CNC program

9. Router Upgrade.. 88

10. Metal Frame Upgrade... 91

11. Arduino GRBL Upgrade...104

12. Laser Engraving Option...113

Bibliography.. 116

Background

You might want to ask me "Why did I write another book on building a CNC Machine?" Here are some reasons for writing this book.

1. Most of the other books are written on the idea that you will need a CNC machine to make a CNC machine. The CNC machine in this book will be made with NO special tools! All you will need is a drill press, drill bits, Jig saw and/or band saw, and optionally a 6-32 and/or 8-32 tap. All holes are 1/2 inch in diameter or smaller.

2. Most CNC plans are for a $1000 or even more expensive machine. My CNC can be built for as little as $300. It can be built for even less than that if you have a large parts collection like mine. Some of the parts I used were even purchased at a yard sale. Many of the parts for this machine can be purchased at your local Lowes or Tractor Supply store.

3. Most CNC plans are for a very specific machine in their design, but my design has a lot of variables that you can use to save money or to make a better machine. You can pick and choose what parts you want to use.

4. Most CNC machines are designed for one specific application. Mine design can be used as a router, a mill, a 3D printer, a laser engraver, and for many other applications. I will eventually include plans for some of these different setups or applications.

5. Many of the parts used in the X, Y and Z axis are basically identical. This is so that when you are making this CNC it is a matter of reusing the same or similar parts throughout.

Design specifications were for a 12 inch by 12 inch by 6 inch work area. Those specifications have not yet been achieved, but it is better to plan high and miss the mark than to not have any goals and hit the mark. Currently it can handle something as big as 11 inches by 8 inches by 4 inches.

Back when I wrote my book on "Arduino Robotics Projects" I came across some 3D printers and CNC mills that used an Arduino as part of their hardware. These projects were a little too advanced for that book, so I thought I would write another book to cover some of these more advanced projects.

CNC stands for "Computer Numerical Control" and refers to a machine that can be controlled with numbers from a computer. There are many standard CNC commands that range from Go to X, Y, and Z coordinates to changing the drilling bit to turning on water cooling. There are many very thick books that cover the many CNC commands in detail. However in this book we will be dealing with just the simplest of CNC commands in order to "Keep it Simple Stupid".

Although I had worked on and repaired CNC machines for many years, and have even wrote some CNC programs for them years ago, this book will mostly be a learning project for me. That is to say that I have a lot to learn in writing this book. I have never built a CNC machine before. Most of my other books are written from my memory and from years of using and teaching electronics. This book assumes that you have read some of my earlier books like the "Arduino Robotics Projects" book where I explained how stepper motors are wired up and how they work.

Making your own CNC router and or 3D printer is a very complex project. It can also be very expensive as these devices can vary from $300 to well over $1000 to build. We will try to start with a "low cost" machine made from surplus parts and then add better parts to make it bigger and more powerful as we go along.

Before you attempt to build your own CNC machine I highly recommend that you look at every design that you can find on the Internet to get some ideas. I even downloaded dozens of pictures and looked through them many times. Try to pick out what features that you like and want to include in your own CNC machine.

Just to make sure that we are all on the same page, the "X" axis moves from the right to the left, the "Y" axis moves from front to back and the "Z" axis moves up and down. Some CNC devices have additional "axis" such as head rotation, filament feed rate, etc.

Chapter 1

Design Concepts

At one time I thought that making an "inexpensive" CNC machine was just a matter of taking a couple of old scanners and turning them into a CNC controlled machine. However the positioning mechanism that is found in most scanners depends heavily on gravity in order to work. So a scanner cannot be used for the top as in "X" and "Z" axis devices. The top assembly of a 3D printer is usually called the "Gantry".

Once I had realized that the scanner could not be used for the "X" and "Z" axis, I thought of using an old printer's mechanism because a printer's mechanics are designed to work while it is mounted with a top to bottom orientation instead of laying flat. So replacing the scanner with a printer might work something like what is shown in the next illustration. There are people who have actually made this kind of setup work.

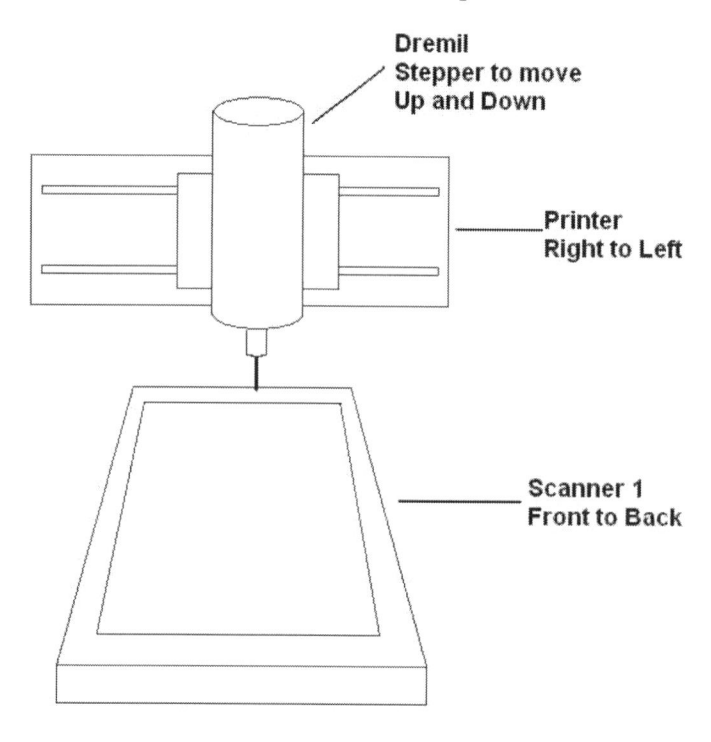

However, almost all printers only run with one guiding shaft, so even when you are using a printer mechanism it is still not completely stable. Also the guiding rods or shafts in most printers vary widely in diameter so it is not that easy to find bearings that will match it perfectly.

I ran into the shaft size issue when I used the shaft from a HP 450 series plotter that was cut in half for the X axis. I measured it as about .375 inches in diameter or 3/8 of an inch. I bought some sleeve bearings for the shaft that were also 3/8 of an inch inside diameter. However they would not fit on the shaft because the shaft actually varied from .375 to .378 inches in diameter! I eventually found some sleeve bearings that were .380 in diameter. They worked on the shafts from that printer so all was not lost.

For the "Z" axis the head positioning mechanism from an older five inch floppy disk drive or CD Rom drive might just work. However that would only give a few inches of travel. Most of the projects we will be working with do not need a large range of motion in the "Z" axis. Also a belt driven mechanism will not work that well for the "Z" axis because it will have to work against gravity and against a drill bit that can sometimes get stuck in the material. So a lead screw mechanism will be required for the Z axis for any milling.

One of the many questions we need to determine is to use lead screw drive or belt drive? A belt drive mechanism works for a scanner or a printer because they always work while traveling in the same direction. For a CNC milling machine, to always have to travel in the same direction can be a serious limitation. This could really be a problem when trying to cut a circle in a circuit board.

Most of the better built CNC machines use what is called a "leadscrew" or "ball screw" type of drive mechanism. Basically a large screw is attached to the stepper motor. A "nut" that rides on the screw is connected to the platform to move it. At the far end of the "lead screw" there needs to be a bearing. Now I need to point out that the threads of a lead screw drive are rounded or almost "U" shaped while the threads of a normal screw are triangle or more like a "V" in shape. The two screw types are not at all compatible with each other. It is best to buy a leadscrew that comes with the matching nut to guarantee their compatibility.

This next diagram compares some of the common drive screw threads.

Common CNC Drive Threads

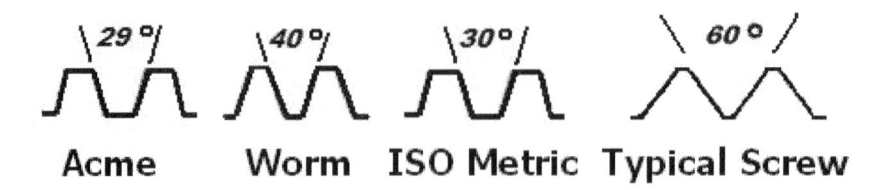

\29°/   \40°/   \30°/   \60°/

Acme   Worm   ISO Metric   Typical Screw

Here is a picture of a 17 inch Teflon coated lead screw drive mechanism. The "Nut" actually has two "nuts" with a spring in between them to keep tension on the nuts and thus reduce the amount of "play". The leadscrew drive in this picture was used for the X and Y axis.

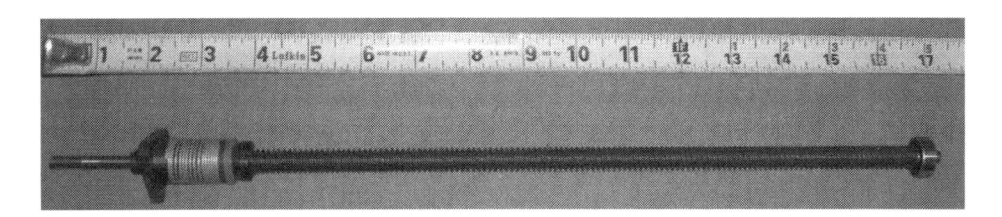

Two guide rods are also used to stabilize the platform and to keep it from rotating. The rods have sliding bearings that connect them to the moving platform. The bigger the stabilizing rods the more stable the table and hence the better the results will be. I tried to use 3/8 inch rods in the X axis and in the center of the travel there was nearly 1/10 an inch of play in the mechanism. To provide 18 inches of travel, 1/2 inch rods give much more stability.

Professional CNC machines have rods that are attached to reinforcing metal in the back or the bottom. These rods are called "Continuously Supported" Rods. The bearings then have to be "C" shaped in order to fit around the supports on the back side of the rods.

The next diagram is of how to set it up a lead screw type of drive mechanism and make it work. A better setup is to have a matching bearing at each end of the drive mechanism and then a flexible coupling to the stepper motor. In my simplified design the bearings in the stepper motor provide the bearings on one end. However these bearings are very small and are only designed for light duty.

The letters "SS" in the drawing refer to "Stainless Steel". Stainless Steel is used because regular steel will rust and that rust would render the machine almost worthless. Zinc coated steel is almost as good as stainless steel as long as the coating is not scraped off over time by some sort of mechanical action.

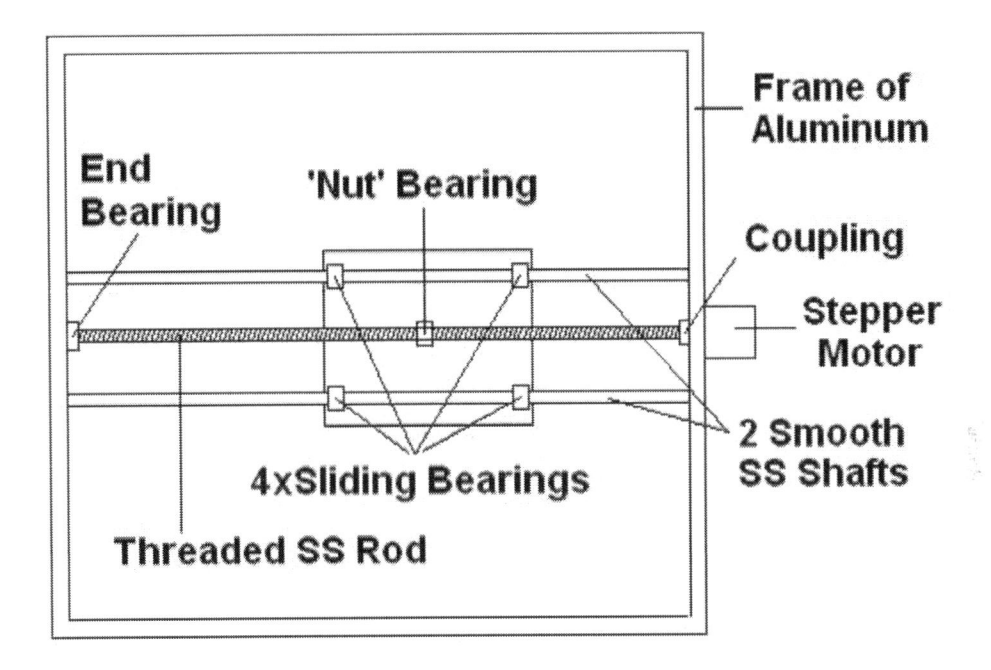

The machine size of 18" by 18" in my machine can be scaled up or down to any needed size. The frame can be made out of aluminum plywood or Plexiglas. The frame height has a limitation in that the stepper motor needs at least a 2.5" wide mounting surface. The stepper motor can be turned 45 degrees and then mounted in place by only two screws. However the use of all four mounting screws is greatly preferred in order to make the machine as stable as possible.

Another important question to answer is if the entire gantry should move up and down or just the tool should move up and down. If the entire gantry moves, then much taller objects can be worked with. If the gantry does not move up and down, then the maximum height of the object on the table is limited by the height of the gantry above the platform.

To move the gantry up and down two matched stepper motors are usually used. One stepper motor is located on each side of the gantry. The problem with this design is in keeping both sides at exactly the same height. It looks something like what you see in this next drawing.

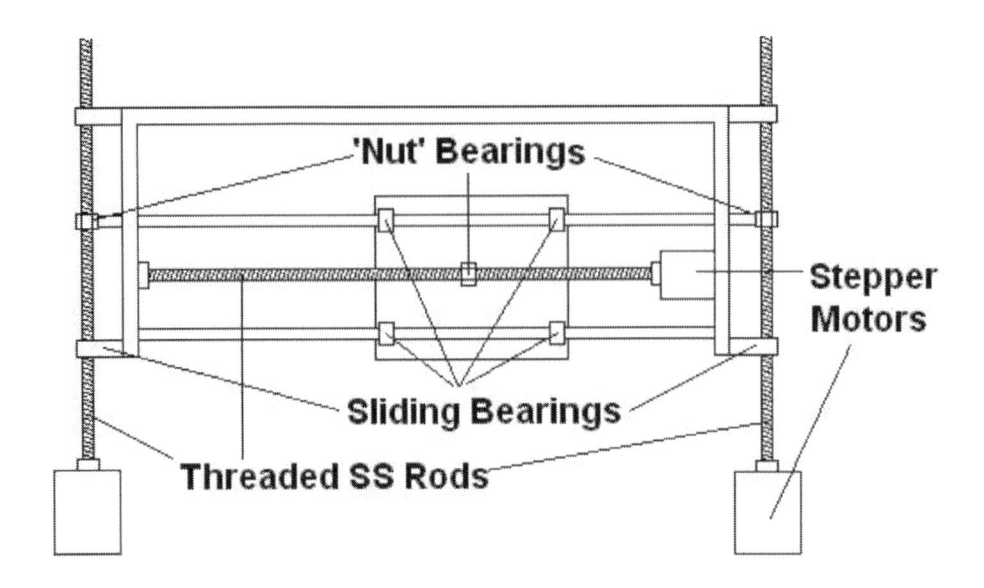

Another design option is to have a fixed platform and then have the gantry move forward and backwards. This can be done by using two matched stepper motors. One stepper is located on each side of the platform. You can also move the gantry in the Y axis with only one stepper motor by having it mounted underneath of the platform and then use a long bar that goes out and connects to the gantry on each side of the platform.

The advantages of having a fixed platform are many, including having more range in the Y axis, but doing so complicates the design and the construction of the CNC machine. A moving gantry also requires a design where all three axes do not work the same as each of the other axis. So to keep the design simple I used a moving platform and a fixed gantry design.

For a starter machine we could use used guide rods from old printers and belt drive systems removed from old scanners. For the "Y" axis we will use two 18" long 1/2" diameter rods removed from wide carriage HP ink jet printers. For the "X" axis we will use a 3' by 1/2" rod removed from an old HP Plotter then cut in half. The belt drives were salvaged from HP ScanJet 5P scanners but they did not last very long. I guess it is good that there are lots of junk HP printers and scanners lying around?

My first complication came when I tried to put the 3/8 inch bearings on the rods removed from a plotter. It turned out that those rods varied in diameter between .376" and .378" inches but the bearings were .375 inch in diameter! So those bearings just simply did not work. I tried enlarging them and even making the rods smaller with emery cloth but that was a lot of work. Then later on I purchased some 1/2 inch rods and made everything 1/2 inch in size instead. It simplifies the design when all the rods and bearings are the same size.

Let's cover some of the more common sizes of guide rods and bearings. By the way, the cheapest solution is to buy 8, 10, 12, or 16 mm rods and bearings to match. You can also use metric rod holders with English sized rods, that is exactly what I did. The metric rods and bearings will likely cost you much less than half of the price of similar English sized rods and bearings!

Here is a very basic English to Metric conversion table. These are only the values that I use frequently.

English	Metric mm	Decimal English
1/4"	6.35 mm	.250" (Close to 6mm)
5/16"	7.9375 mm	.3125" (Very Close to 8mm)
3/8"	9.525 mm	.375" (Close to 10mm)
7/16"	11.1125 mm	.4375
1/2"	12.7 mm	.500" (Close to 13mm)
5/8"	15.875 mm	.625" (Close to 16mm)
3/4"	19.05 mm	.750" (Very close to 19mm)
7/8"	22.225 mm	.875" (Close to 22mm)

I dug through all of the garbage that I have collected in the shed and came out with two 18 inch by 18 inch by 3/8 inch thick pieces of Plexiglas. There were several other smaller pieces of Plexiglas in there as well. Some of the smaller pieces were used for the Z axis. One of the larger pieces had been used to build a dual 812 vacuum tube Tesla coil at one time so there were several holes already drilled in it.

You can start by cutting a 24 inch by 24 inch sheet of Plexiglas or plywood into the pieces needed to build the CNC machine. This next diagram makes an 18 inch square base, two 18 inches by three inch ends and 12 inches by two six inch sides for the gantry. If you are using parts from old printers you might want to start by making a slightly smaller CNC machine as most "normal" printer parts are not 18 inch long.

Coming up next is the 24" by 24" plywood or Plexiglas cutting plan diagram.

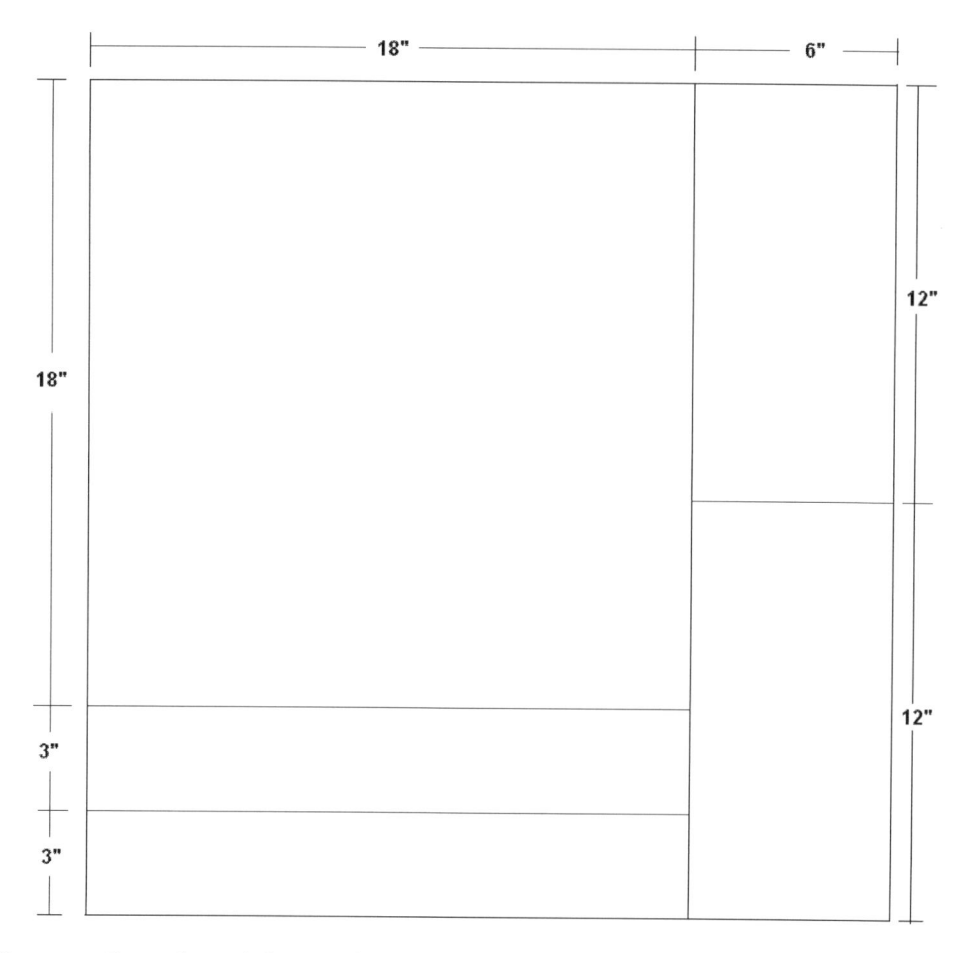

Later on I purchased four 24 inches by 8 inch pieces of Plexiglas at a yard sale. They were also used in building the CNC machine. They were purchased for only $1 each! I used one of them to make both the Y axis and Z axis platforms. The Y platform was made out of a piece of 3/4 inch plywood before I replaced it with the Plexiglas. I still use that plywood as a "crash guard".

Since the pieces of Plexiglas that I started with were both 18 inches by 18 inches here is how I cut one of them up. Between the two pieces of Plexiglas I had the same parts that were shown with the 24 inch by 24 inch design that was shown above.

```
┌─────────────┬─────────────┬─────────────┐
│             │             │             │
│             │             │             │
│             │             │             │
│ Left gantry │ X Platform  │ Right gantry│
│ 6" by 12"   │ 6" by 12"   │ 6" by 12"   │
│             │             │             │
│             │             │             │
│             │             │             │
├─────────────┴─────────────┴─────────────┤
│            Y Axis End                   │
│            18" by 3"                    │
├─────────────────────────────────────────┤
│            Y Axis End                   │
│            18" by 3"                    │
└─────────────────────────────────────────┘
```

Up next there is a picture of the freshly cut pieces of plastic lying on the table. There are some smaller two inches by seven inch pieces of plastic in this picture that were used to make the ends of the Z axis.

I had also collected the stepper motors from some scanners and some angle brackets that are visible in the picture as well.

The masking tape was there for use in marking where to drill the holes in the plastic. New sheets of plastic come with a film on them that is easy to mark up and that works much like the masking tape does.

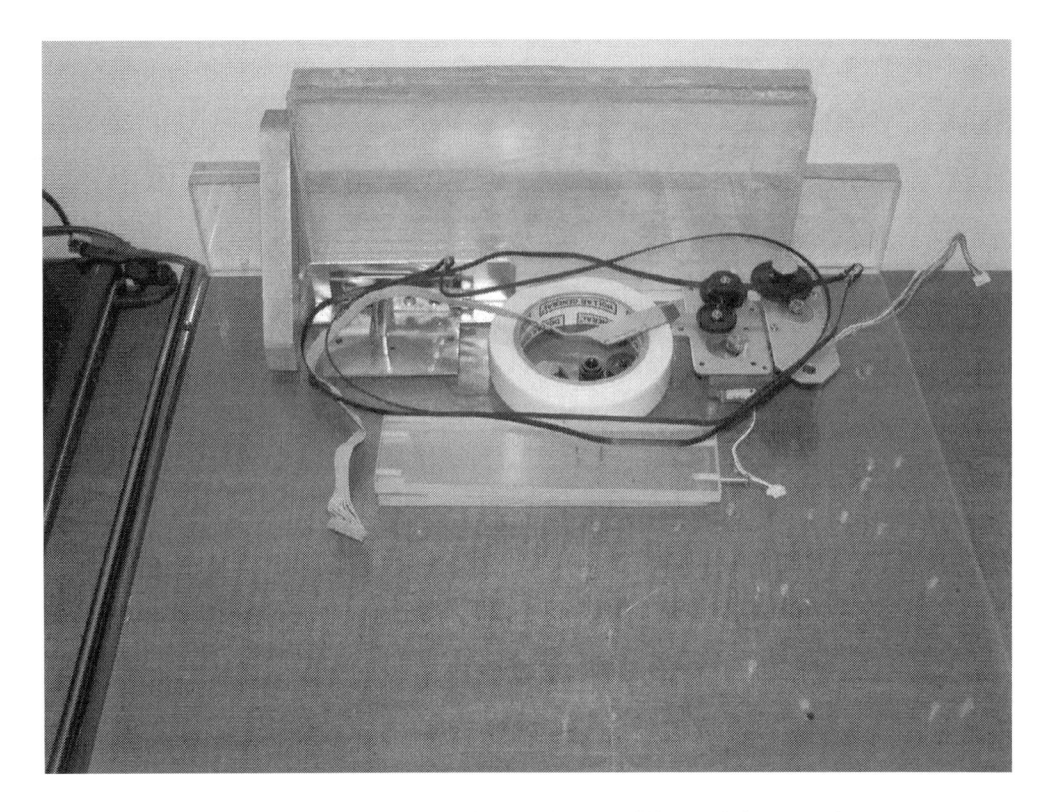

If you are making your CNC out of metal you will need four 3" by 1" C channel pieces that are 18 inches long. Then you will need two pieces 12 inches long and two pieces that are 6 inches long. The X, Y and Z platforms would all be the same size as they are for plywood or Plexiglas design.

To drill the metal parts first cover them in masking tape. Then use a sharp pencil to mark them and double check all of your measurements before beginning to drill them. Use a center punch to make sure that the holes are at the right spots and drill a small hole first. Then drill it out to the correct size. Almost all of the holes are clearance holes for the 10-32 screws except for the motor shaft holes.

You could also use 8-32 screws if you prefer. There are also a few larger holes for the 1/4 inch shafts of the stepper motors that need to be slightly larger to allow the motor shafts room to spare. This next picture is of the metal parts cut and drilled. If you look carefully you can see that I missed a few holes here and there.

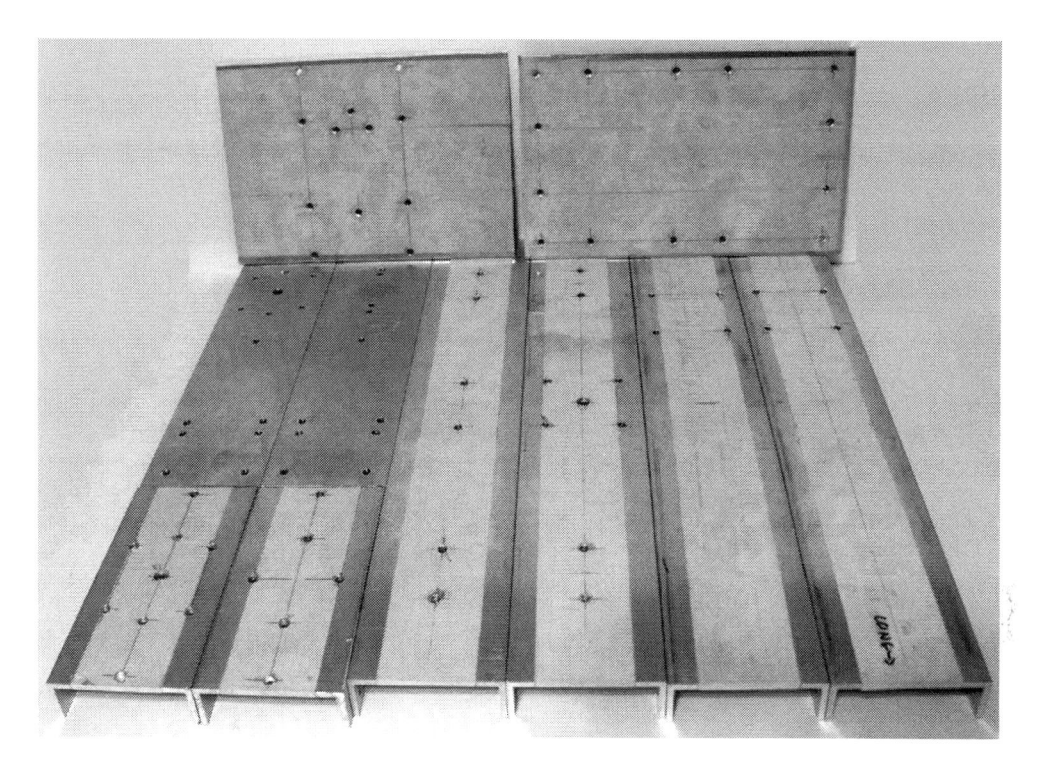

The next drawing is how I had envisioned that the CNC machine might eventually be put together. This drawing is oriented as though you were looking down from above at the base or at the Y axis of the CNC. The Y axis guide rails are spaced 12 inches apart and they are spaced in three inches from each side.

The first design had a horizontally mounted stepper motor and drive belt but they were replaced with a leadscrew setup after just one test run. There were just too many problems with trying to use a belt drive system. At the opposite end of the leadscrew from the stepper motor there is a bearing and fastening that bearing in place can be a bit tricky.

You can make your own bearing mounting bracket out of Plexiglas or plywood by using a jigsaw to cut a hole big enough to fit tightly over the bearing. In the metal version there is a metal bracket that holds the bearings in place.

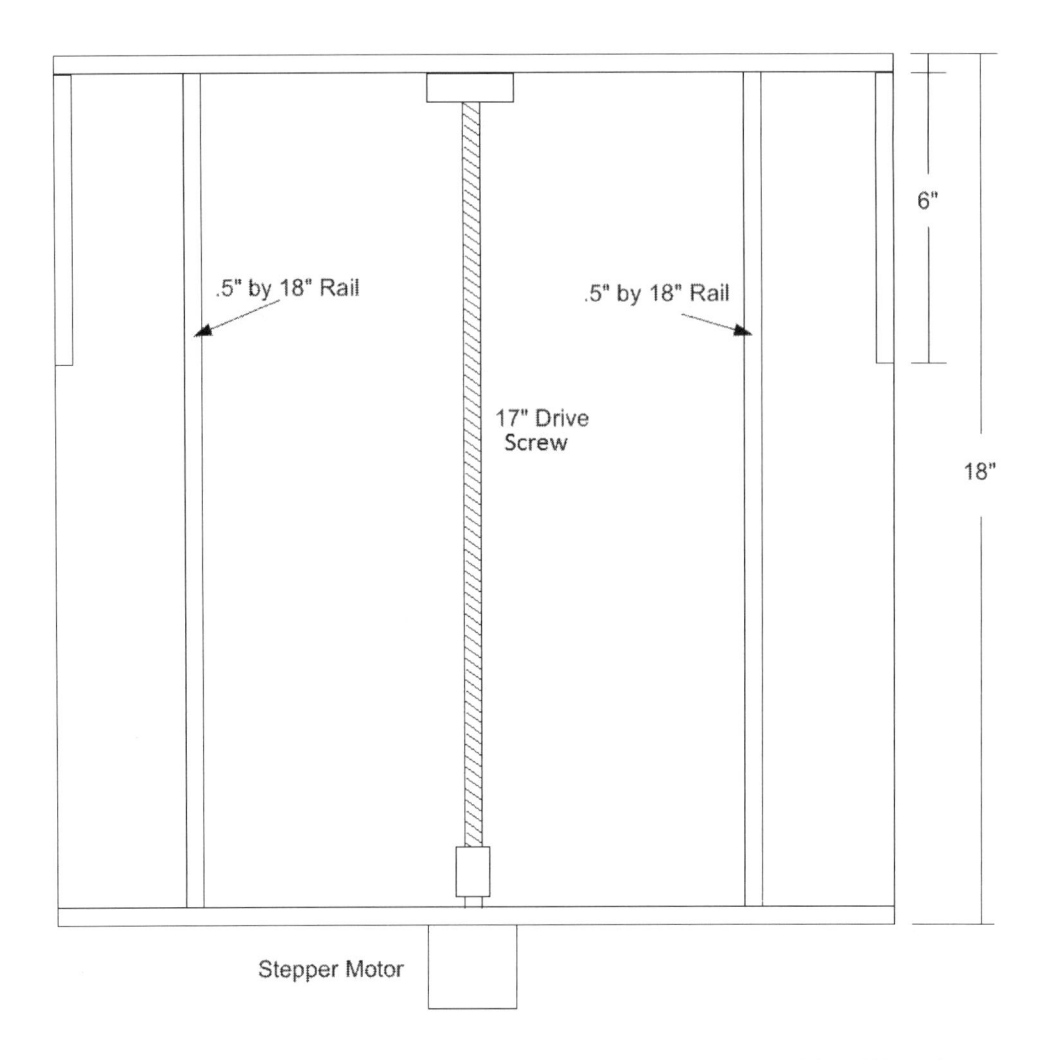

The gantry or X axis was then designed to end up look something like what is shown in this next drawing. The drawing is from the viewpoint of standing in front of the CNC looking straight at it with the Y axis at the bottom front of the picture.

The X axis support rods are located at one inch from the top and at five inches from the top and they are spaced 1.5 inches back from the front edge. This arrangement was used to keep the design and numbers similar to what was used in making the Y axis. The original 3/8 inch rails were replaced with 1/2 inch rails to make the machine more stable and more consistent throughout.

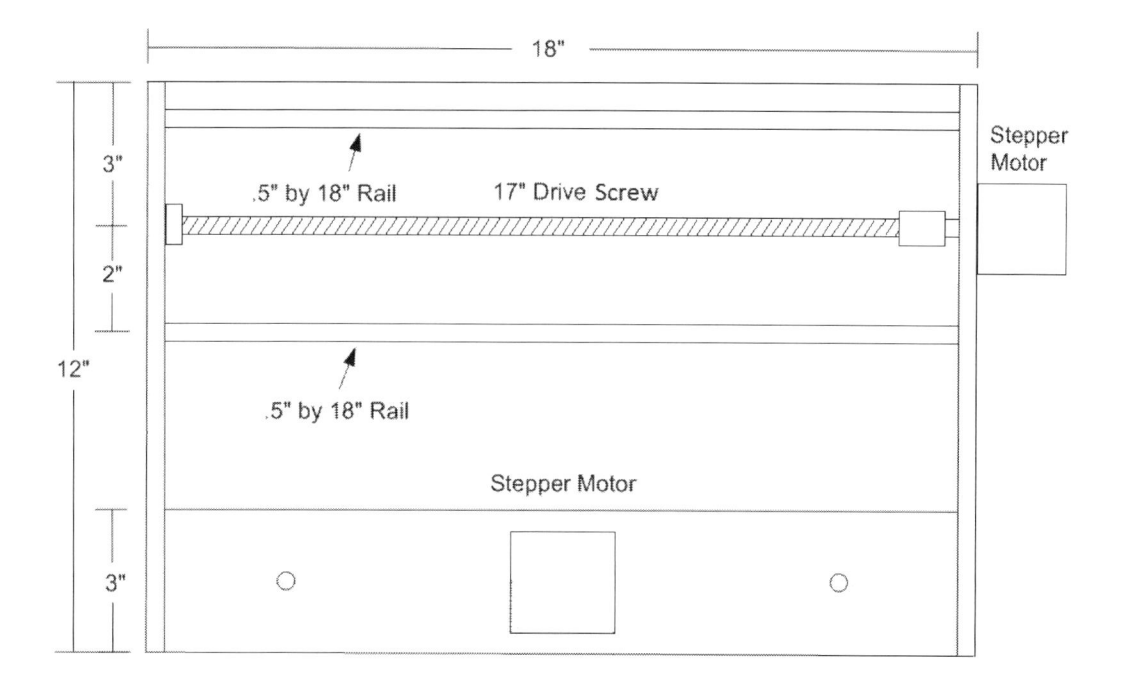

[Diagram: Top-down view of CNC frame. Overall dimensions 18" wide × 12" tall. Top section 3", middle 2", bottom 3". Shows .5" by 18" Rail, 17" Drive Screw, another .5" by 18" Rail, Stepper Motor on the right side, and Stepper Motor at the bottom center.]

I decided to use metal angle brackets to fasten the plastic pieces together. Drilling and taping holes in the plastic would have looked much better. I was afraid that drilling too close to the edges of the plastic posed a risk that it would break.

Speaking of drilling, if at all possible USE A DRILL PRESS! When you are drilling plastic the drill bit gets hot and it then melts the hole sideways in the plastic unless you are using a drill press. To drill larger holes like the 1/2 inch holes, drill a 1/8 inch pilot hole first. Then drill the larger size hole from both sides so that they meet in the middle. This will reduce the chance of pieces of plastic breaking out. Also for the sides of the gantry, front and back of the Y axis, and the top and bottom of the Z axis, you can clamp the two sides together to drill the pilot holes so that the position of the pilot holes will match exactly on each piece.

Here is a picture of the plastic pieces put together with some of the rods and stepper motors installed. There are also some holes drilled in the plastic to mount some NEMA 23 stepper motors in a future application. I will have to talk to the Robo-raptor as he was nibbling on the upper right corner of the CNC. That is because he is jealous that I am no longer playing with him and am spending all of my free time building the CNC machine. This setup was an attempt at using belt drive.

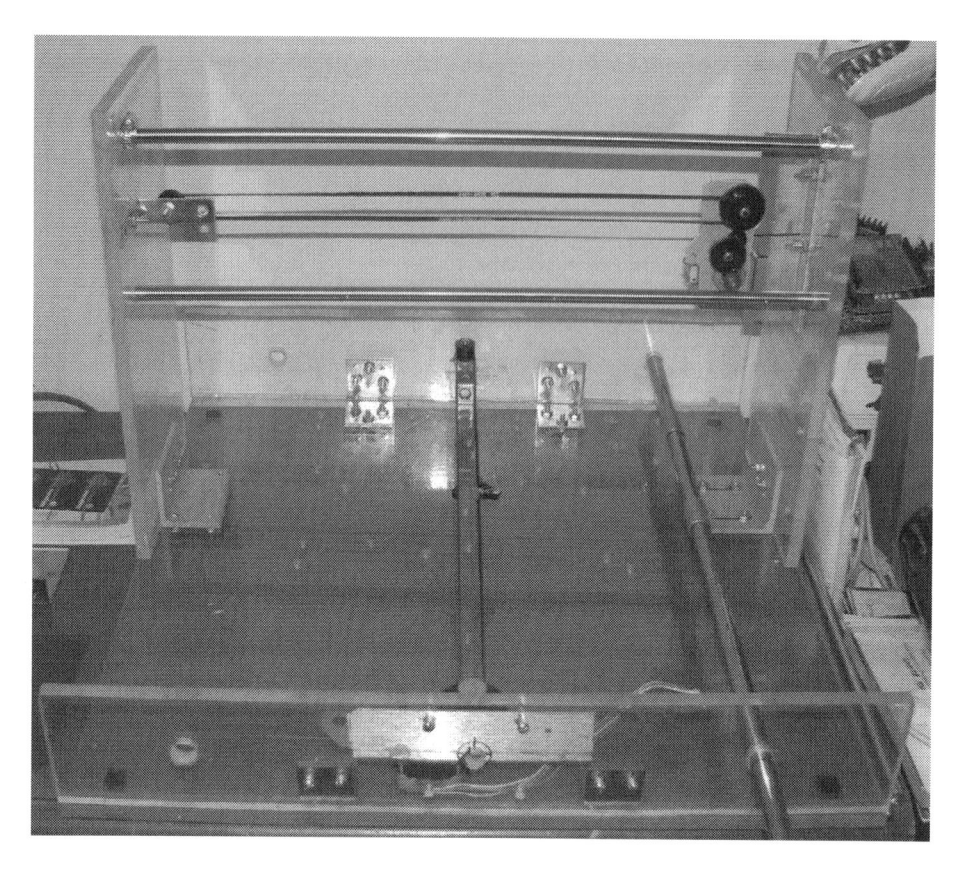

The stepper motor mounting bracket for the belt drive mechanism was a bit cobbled together but it worked. The problems were with the belt tensioning spring. It gives easily when under pressure. The weight of the platform alone created too much pressure for the belt drive.

This next picture is from the start of assembling the metal version. I had to use a square to make sure that the base was square and that the gantry sides were perpendicular to the base. Several holes had to be filed to make the screws fit and so that the parts aligned properly. I had the most problems with bolting the corners of the base together.

The metal parts can be cleaned with an emery cloth or with really fine sandpaper to get them to shine like they were actually new. They were some scrap pieces that I had salvaged from a recycle bin.

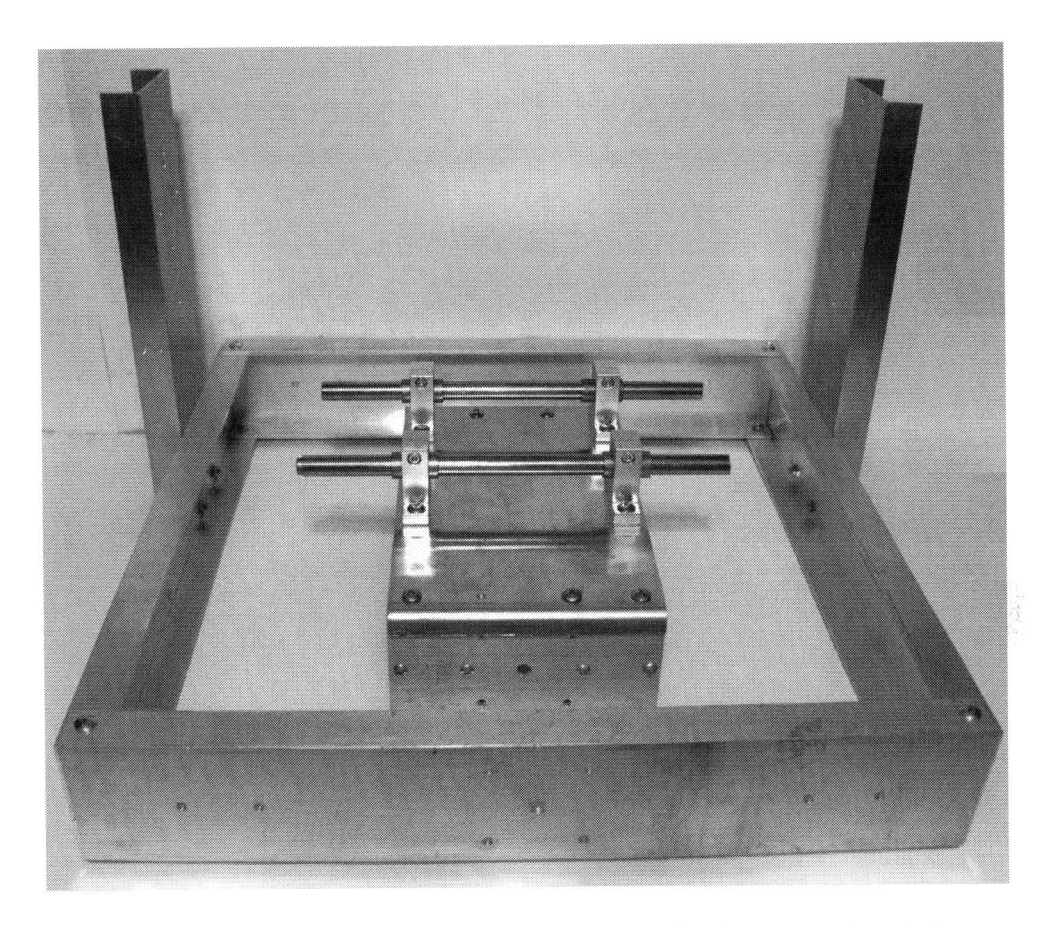

My first test run of the machine was what appeared to be a complete failure. First the movement bound up to where I could not even move the platform around by hand. Then the stepper did not move the platform and the belt just bunched up and slipped. Then the motor driver got so hot that it could almost burn your fingers if you touched it.

The binding problem was from the holes in the front and back of the Y axis not matching up exactly. I had to file one of the holes slightly larger to the right and then the binding problem was fixed. The problems with the belt drive were fixed by changing to using a ball screw type of drive mechanism instead.

The stepper driver was getting so hot because I was trying to drive the motor way too fast, so only one out of about every ten pulses were actually doing anything. I had to add a delay between each of the step commands to get the motor to move smoother and faster and run cooler.

For the second test run the machines was set up to work like a giant etch-a-sketch. Some of the problems encountered included that there was no tool mount on the Z axis yet so the pen wandered all over as it was not held in place. Also the Y axis slipped once again.

To fix the Y axis I changed the sleeve used to adapt the leadscrew shaft up to 1/4 of an inch to fit the stepper motor shaft coupler. Instead of it completely surrounding the shaft I made it into a "C" so one of the set screws could go directly into the shaft. I also made a different bearing holder that would not allow the leadscrew to move away from the shaft coupler. You can see the adapter sleeve in this picture.

For the next test I changed the stepper controllers and added the home made pendant found later in this book. I was then able to use a magic marker and run a "Flatness" test. This test determines how much the distance from the Z axis to the Y axis platform varies over the surface of the Y axis platform. If the pen becomes crushed or skips then the Y platform is not completely flat.

As you can see in the next picture the front left corner was lower resulting in the pen skipping there.

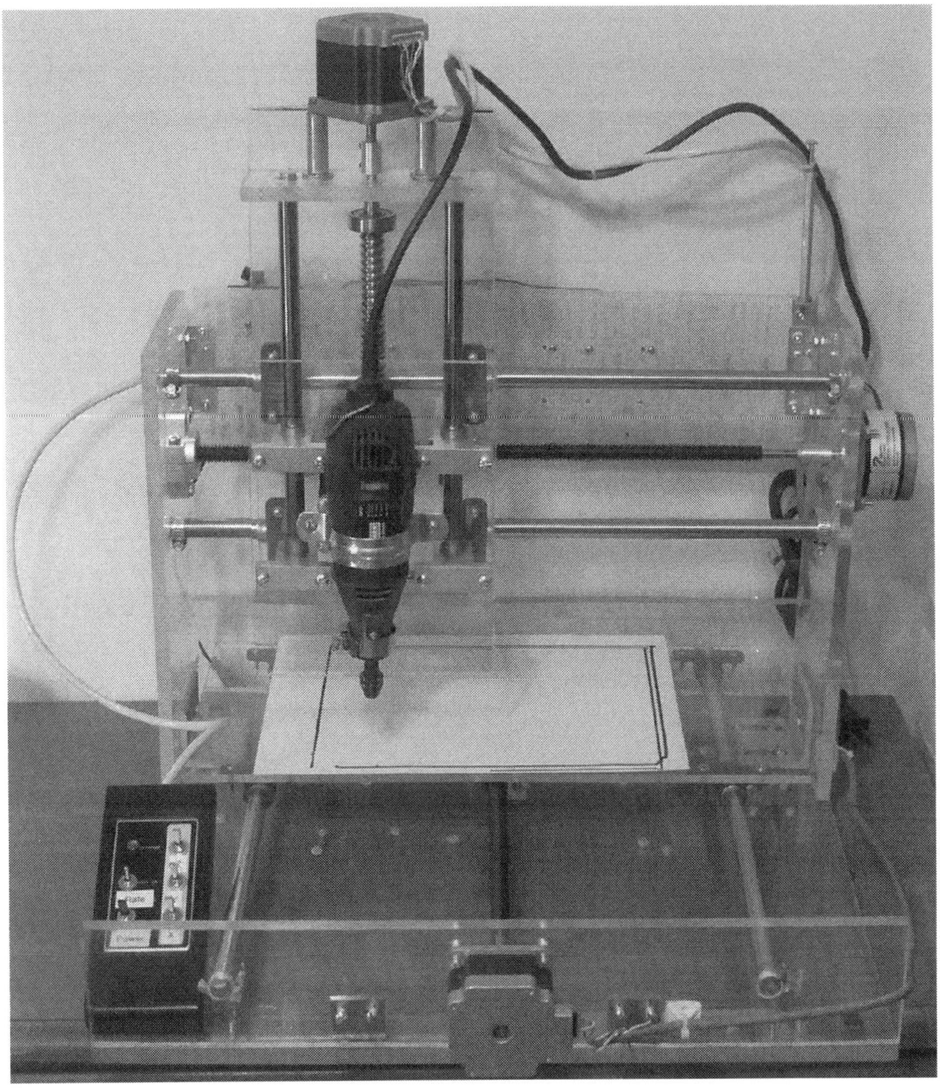

Chapter 2

Materials Used

My budget for this CNC machine was only $100 when I started out. That did not last long as things needed to be improved on or to be replaced. When I first totaled up the amount that was spent on this machine it was shocking! I have broken down the parts to what were needed parts, what parts were optional and what turned out to be wasted money.

Here is the parts list with their price and source:

1 x Z axis lead screw drive, with bearings	$22	(eBay)
3 x NEMA 23 stepper motors	$41	(eBay)
2 x X and Y axis lead screw drive	$34.50	(eBay)
2 x X and Y axis drive bearings R4A-2RS	$5.60	(eBay)
2 x X and Y axis SHF20 Bearing supports	$8	(eBay)
4 x 1/2 inch by 18 inch X and Y axis rods	$12 TS or $40 eBay	
(The 1/2 inch by 10 inch rods in the Z axis were from a printer)		
3 x 1/4 inch shaft couplers	$9.50	(eBay)
12 x 1/2 inch brass sleeves	$36	(Lowes)
10 x 1/2 inch plastic sleeve pipe holders	$6	(Lowes)
4 x 16mm metal bearing supports	$18	(eBay)
2 x 20mm Shaft Supports	$8	(eBay)
10 x 1/2 inch Collars	$12	(eBay)
3 x 6560 Stepper Controllers	$30	(eBay)
1 x 12 Volt 5 Amp AC adapter	$10	(eBay)
Several 6-32 and 8-32 screw assortments	$4	(Lowes)
10 x 6-32 flat head screws for the Y axis	$2	(Lowes)
8 x 2.5 x 1.0 x 1.0 angle brackets	$24	(Lowes)

TOTAL   Approx. $300

NEEDED PARTS

On the next few pages there are some pictures of the items that were purchased on eBay, Lowes or Tractor supply, along with a short description or explanation about it.

The "lead screws" were advertised as "12 inches" but they are actually over 17 inches long. There is only about 14 inches of travel and the "nut" assembly is two inches long leaving only 12 inches of useful travel. Another problem with them was that the bearing ends were 1/4 inch in diameter but the end that the stepper connects to was a little bit smaller than 1/4 of an inch. I used an aluminum sleeve to adapt it back up to 1/4 an inch so they fit the 1/4 inch shaft couplers. The aluminum sleeve was about as thick as two sheets of paper.

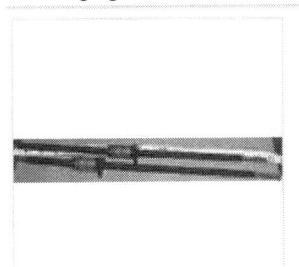

two Leadscrews Lead Screw 12" 300mm teflon
(251470731721)
Sale date: 03/16/14
Tracking number: 9405509699938294071942
Delivered: Wed. Mar. 19

$25.49
+ $8.99 shipping

I purchased one of these 1/2 inch by 18 inch linear shaft from XYB Bearing on eBay but there were two of them in the box. They are really high quality.

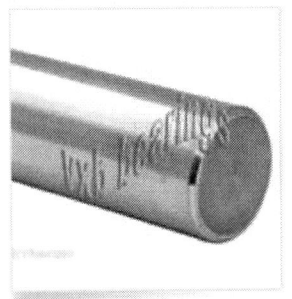

NB Linear Systems PC8-18" 1/2" Pre-Cut Slide Shaft 18" in... (131135542528)
Sale date: 03/15/14
Tracking number: 1Z6191E53974348582
Estimated Delivery: Thu. Mar. 20

$19.95
Free shipping

To save some money you can get a zinc coated 1/2 inch by 36 inch rod for $6 at tractor supply. That rod is not as smooth nor is it exactly 1/2 inch in diameter, but it will work in a pinch. Here is a picture of the rod from Tractor Supply. When cutting it in half, be sure to cut 1/4 of the way through it all the way around it then cut it the rest of the way in half. If you do not do it that way the bearings might not fit on it. One end of the rod I purchased had that problem from the factory when I bought it.

National Hardware® 4005BC 1/2 in. x 36 in. Smooth Rod, Zinc (Blue)

$5.79

Be the first to review this product
Write a review

✓ In Stores Only - Check Availability

14221   Check Store Availability

SKU 3504558

Metal | Brand : National Hardware® | Product Type : Round | Product Width : 1/2 in. | Product Surface : Smooth | Product Length : 36 in. | Product Height : 1/2 in. | Product Thickness : 1/2 in. | Material : Steel | Product Finish : Zinc
▸ More

The Z axis leadscrew used 3/8 inch bearings. However the outside of those bearings was about an inch and a quarter in diameter! The only thing I could find to fit that size was a capacitor mounting strap. The total leadscrew length was 10.5 inches the screw travel was nine inches long. The bearing area was two inches long leaving seven inches of useful travel.

ACME leadscrew 16mmx160mm travel, Anti-backlash, matched bearings (291089322995)
Sale date: 03/04/14
Tracking number: 9405509699938268992020
Delivered: Mon. Mar. 10

$10.50
+ $11.30 shipping

There were problems with the stepper motors too. First of all I paid a little bit too much for them. You can get three NEMA 23 stepper motors for about $35 and sometimes that even includes the shipping. Of course I had bought 6 steppers motors but only three of them were the needed NEMA 23 size.

Another problem with the stepper motors was that they had metal gears fastened on their shafts. I tried everything to get the gears off in one piece but finally used a hack saw to cut down one side of the shaft cutting the gears in half and making one side of the shaft flat in a few spots. This was very hard to do without damaging the stepper motor.

A third problem with these stepper motors was that two of the stepper motors did not have their plugs or wires included with them. I had to solder wires directly on to one stepper motor because the pin spacing did not match any connector that I could find.

LOT 6 STEPPER MOTOR STEPPING STEP GEARHEAD JAPAN SERVO (161217410375)
Sale date: 02/13/14
Tracking number: 9405510200828084313672

$27.00
+ $13.95 shipping

Up next is the SHF20 shaft supports that I bought to hold the X and Y leadscrew bearings in place. They should have been 19 mm as that is closer to the needed 3/4 inch diameter size of the bearings. Also these were so big that they got in the way and reduced the travel in the axis by almost one inch.

2pcs SHF20 20mm Linear Rail Shaft Support XYZ Table CNC (400332135626)
Sale date: 04/04/14
Tracking number: LK010262995CN
Estimated Delivery: Tue. Apr. 22 - Thu. May. 8

$7.99
Free shipping

Instead of the SHF20 shaft supports you could use 1 inch Capacitor clamps. Because the X and Y bearings are only 3/4 of an inch in size it will take several layers of electrical tape or something else to adapt them to the size of these clamps. These clamps did work well for the Z axis bearings.

4pcs 25mm 1" Film Electrolytic Can Capacitor metal Clamp ... (261257466935)
Sale date: 05/16/14
Tracking number: LK058182182CN
Estimated Delivery: Wed. May. 28 - Thu. Jun. 12

$4.49
Free shipping

I also had to purchase the ball bearings for the ends of the X and Y axis lead screws. There are two different bearing outer sizes available and I chose to use the larger ones as they should be more durable.

	R4A-2RS Premium Sealed Bearing, 1/4" x 3/4" x 9/32" R4A R... (370922175596) Sale date: 04/04/14 Tracking number: 9400110200881141836948 Estimated Delivery: Thu. Apr. 10	$2.79 Free shipping

The 1/4 inch shaft couplings were a little smaller than what I though they would be. As a result the ends of the stepper motors had to be very close to the ends of the lead screws. I was hoping that they were a little longer to accommodate the slightly short lead screws that I am using. They also should have been flexible ones as there are slight fluctuations that come from the shaft size differences.

	1/4" SHAFT COUPLERS or COUPLINGS - LOT OF 3 PIECES (251384433455) Sale date: 04/04/14 Tracking number: 9400109699938347262577 Estimated Delivery: Wed. Apr. 9	$9.48 Free shipping

Eventually I purchased two of these TB6560 stepper controllers. They work great! The machine might have never worked properly without them. Not all TB6560 controllers are the same. I purchased one from another seller and it barely worked at all. On closer examination some of the parts values were not even correct!

	TB6560 3A Driver Board CNC Router Single 1 Axis Controlle... (200976756349) Quantity: 2 Sale date: 04/09/14 Tracking number: LK015994895CN Estimated Delivery: Sat. Apr. 19 - Mon. Apr. 28	$19.10 Free shipping

For the Z axis there were four 16 mm shaft supports that fit over the sleeve bearings. These shaft supports should really be used for all three axes.

2x SK16 Size 16mm CNC Linear Rail Shaft Guide Support   MPH (331112396493) $8.90
Quantity: 2   Free shipping
Sale date: 05/27/14
Tracking number: --
Estimated Delivery: Mon. Jun. 16 - Thu. Jul. 17

Some of the parts were from Lowe's or Tractor Supply. The first part is the 1/2 bronze sleeve bearings. You will need a total of 12 of these as there are four needed for each of the three axes.

The Hillman Group 1/2-in x 5/8-in Bronze Standard (SAE) Flat Washer $2.63
Not Yet Rated   Qty.: 1
Item #: 215749   Model #: 882746

- Sleeve bearings are used in many different machines such as automobiles, home appliances and office machinery
- Allows free movement of machine parts

To fasten the bronze sleeves in place I used these 1/2 inch plastic pipe supports as they are slightly larger than 1/2 an inch in diameter. You could replace all of these with the SK16 16mm shaft supports instead.

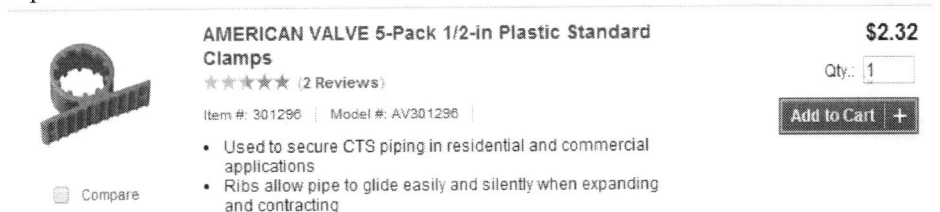

AMERICAN VALVE 5-Pack 1/2-in Plastic Standard Clamps $2.32
★★★★★ (2 Reviews)   Qty.: 1
Item #: 301296   Model #: AV301296

- Used to secure CTS piping in residential and commercial applications
- Ribs allow pipe to glide easily and silently when expanding and contracting
- Split design allows for use after piping has been installed

There are screw and nut assortments that include 6-32 and 8-32 screws such as these. You will likely need more than one kit. All together it takes about 100 screws and 100 nuts to build this CNC machine. About half are 6-32 and half are 8-32 screws.

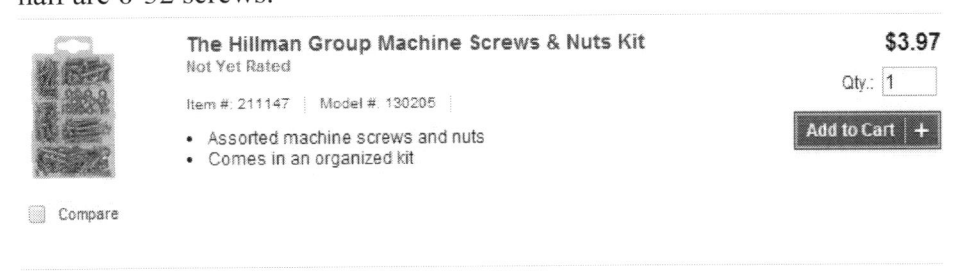

The Hillman Group Machine Screws & Nuts Kit $3.97
Not Yet Rated   Qty.: 1
Item #: 211147   Model #: 130205

- Assorted machine screws and nuts
- Comes in an organized kit

Lowe's also carries many corner braces. I used four 1.5 inch braces like the third one below, and 2 of the 2.5 inch braces like the second one below. For the gantry I used much bigger braces but I have not found any source for getting them. You could likely use four of the top brace for the gantry sides.

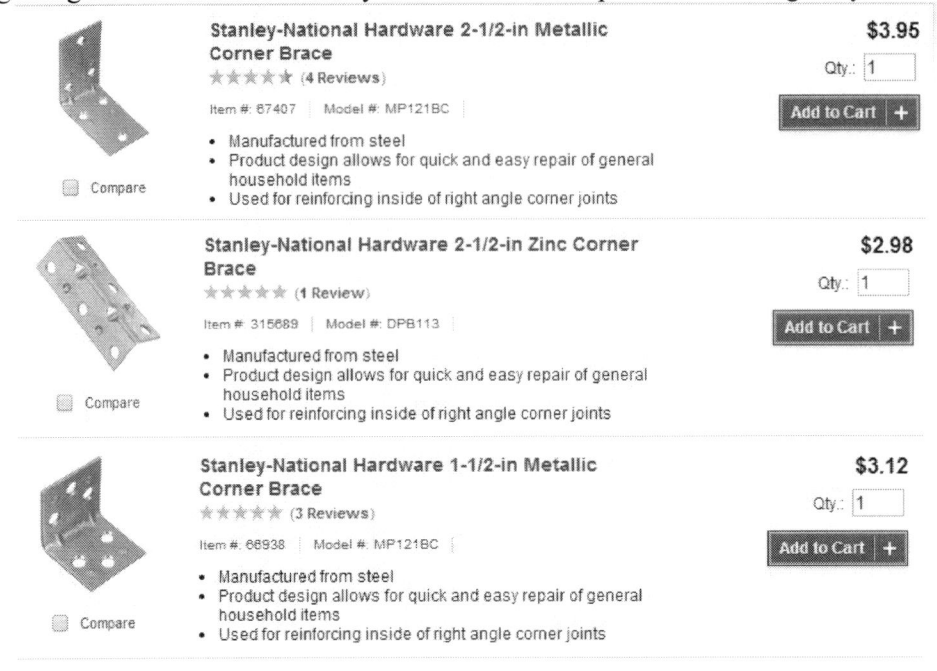

	Stanley-National Hardware 2-1/2-in Metallic Corner Brace	$3.95
	(4 Reviews)	Qty.: 1
	Item #: 67407 Model #: MP121BC	Add to Cart +
	• Manufactured from steel • Product design allows for quick and easy repair of general household items • Used for reinforcing inside of right angle corner joints	

	Stanley-National Hardware 2-1/2-in Zinc Corner Brace	$2.98
	(1 Review)	Qty.: 1
	Item #: 315689 Model #: DPB113	Add to Cart +
	• Manufactured from steel • Product design allows for quick and easy repair of general household items • Used for reinforcing inside of right angle corner joints	

	Stanley-National Hardware 1-1/2-in Metallic Corner Brace	$3.12
	(3 Reviews)	Qty.: 1
	Item #: 66938 Model #: MP121BC	Add to Cart +
	• Manufactured from steel • Product design allows for quick and easy repair of general household items • Used for reinforcing inside of right angle corner joints	

OPTIONAL PARTS

The next few parts are optional as they are not needed to get the CNC to work.

To hold the 1/2 inch shafts in place I used small pipe clamps at first. They are cheap but they do not look very "professional". Then I found these 1/2 inch collars on eBay. They look really professional.

Shaft Collar - 1/2" ZINK - NEW - 10 PACK (261239156015) $13.99
Free shipping
Sale date: 05/27/14
Tracking number: 9400109699938484463424
Estimated Delivery: Mon. Jun. 2

For the pendant I bought a rotary encoder. This is an optional part but it makes it easier to work with so you can manually generate "Steps".

1x 12mm Shaft Rotary Encoder Switches Dia 6MM EC11 A617 HM (181361695762) $0.99
Free shipping
Sale date: 05/19/14
Tracking number: --
Estimated Delivery: Fri. Jun. 6 - Wed. Jul. 9

I also purchased a replacement stepper motor for the Z axis because that stepper likes to sometimes scrape on its cover.

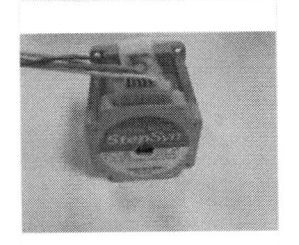

Nema 23 Stepper motor 6-wire unipolar 1.9A 1.8 degree (331111839987) $6.00
+ $9.92 shipping
Sale date: 05/20/14
Tracking number: 9405509699937590346129
Delivered: Fri. May. 23

I purchased this A4988 "Stepstick" stepper driver board in order to test it out. It requires an adapter board to be able to connect it up. My hope was that it would be faster than the current stepper motor driver boards. The interface schematic is in the electronics section of this book.

StepStick Stepper motor driver A4988 A4983 3D Printer dri... (291120901181) $3.05
Sale date: 05/13/14 Free shipping
Tracking number: --
Estimated Delivery: Wed. Jun. 4 - Tue. Jul. 1

Since I had some Plexiglas and bought more of it at a yard sale I did not realize what the normal cost would be to purchase it. I found this listing on Amazon but there are some companies that charge as much as $80 for the Plexiglas.

by Falken Design

24" x 24" - 3/8" Clear Extruded Acrylic Plexiglass Sheet

Be the first to review this item

Price: $38.80 + $11.99 shipping

In stock.
Usually ships within 2 to 3 days.
Ships from and sold by Falken Design Corporation.

Roll over image to zoom in

Specifications for this item

Brand Name	Falken Design
Part Number	acrylic_clear.354_24x24
Material Type	Acryllic
Item Shape	Sheet
Item Weight	11 pounds
Pkg Qty	1
UNSPSC Code	30000000

WASTED MONEY

There were also some things that I bought for this project but either never used them or was unable to get them to work. These items were a waste of my money. I included these to show that there are always things that go wrong so do not feel bad if you buy something and it does not work out.

There are the L297 IC's that are used to interpret the step and direction signals into four phases to go to the L298 power driver IC or to TIP120 power transistors to drive the stepper motors. They require a lot of connections to work. I never got them to work correctly.

5PCS STEPPER MOTOR CTRLR IC ST DIP-20 L297 L297/1 NEW GO... (271413691603)
Sale date: 04/02/14
Tracking number: LK007508512CN
Estimated Delivery: Sat. Apr. 12 - Fri. Apr. 18

$3.20
+ $2.00 shipping

The L298 boards are kind of a mistake as they do not work directly with most CNC equipment. CNC equipment uses "step" and "direction" commands going to the stepper motor controllers. L298 uses phase 1, 2, 3, and 4 as its inputs. The L298's cannot be used without adding some L297's so that the step and direction commands can be used.

Dual H Bridge DC Stepper Motor Drive Controller Board Mod... (181298308755)
Quantity: 2
Sale date: 02/18/14
Tracking number: LK953779211CN
Delivered: Sat. Mar. 1

$5.72
Free shipping

These LM10UU 10mm linear ball bearings were the wrong part altogether. Somehow I thought that the X axis support rods were 10mm in size when they were found to be a little over 3/8 of an inch. Note that 10mm is a little larger than 3/8 of an inch. You might need a micrometer to tell the difference.

10pcs LM10UU 10mm $8.55
10x19x29mm Linear Ball
Bearing Bush Bus... Free shipping
(121170026097)
Sale date: 03/17/14
Tracking number: LK986692994CN
Delivered: Mon. Mar. 31

These 3/8 inch bunting bearings were another mistake. They were supposed to fit the 3/8 inch rods in the X axis but they would not even fit onto the rails. I tried enlarging them using emery cloth and a drill but that took a long time and it roughened the bearing surfaces.

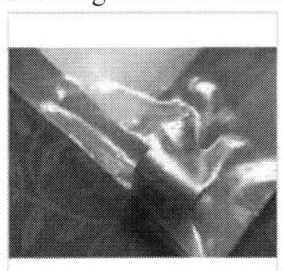

6 Bunting Bearings AA521 $7.99
3/8" x 1/2" OD x 3/4" long
Free shipping
Powde... (321326881883)
Sale date: 03/04/14
Tracking number:
9400109699938263912914
Delivered: Mon. Mar. 10

Chapter 3

Mechanical Design

The stepper motors can be mounted in one of three different ways. One mounting method is to use extra long M4 metric screws if the holes are threaded for them. An alternative method is to use long 6-32 screws with nuts. Be sure to use four thick washers, one for each corner, or eight thin washers to space the motor back to accompany the protruding bearing support that is built into the stepper motor. Either use spacers or you have to make the hole for the motor shaft that is 1.5 inches in diameter. A third mounting method is to re-tap the motor mounting holes with an 8-32 tap and then you can use 8-32 screws to mount the stepper motors.

Here is the typical spacing for a NEMA 23 motor mounting holes. The center hole should be 1.5 inches in diameter. Since you cannot drill a hole that big without special drilling equipment, I used a 1/2 inch hole and then spaced the stepper motor back with two washers at each mounting hole. You can also use a thicker lock washer for the spacer.

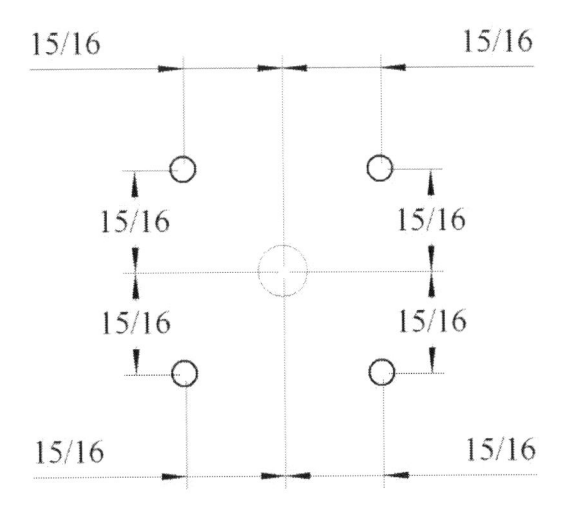

```
        15/16                        15/16
    ─────────────▶        ◀─────────────
                  │  ○        ○  │
                15/16            15/16
                  │                │
                         ◯
                  │                │
                15/16            15/16
                  │  ○        ○  │
    ─────────────▶        ◀─────────────
        15/16                        15/16
```

On the next page there is the mechanical specification for a typical NEMA 23 stepper motor.

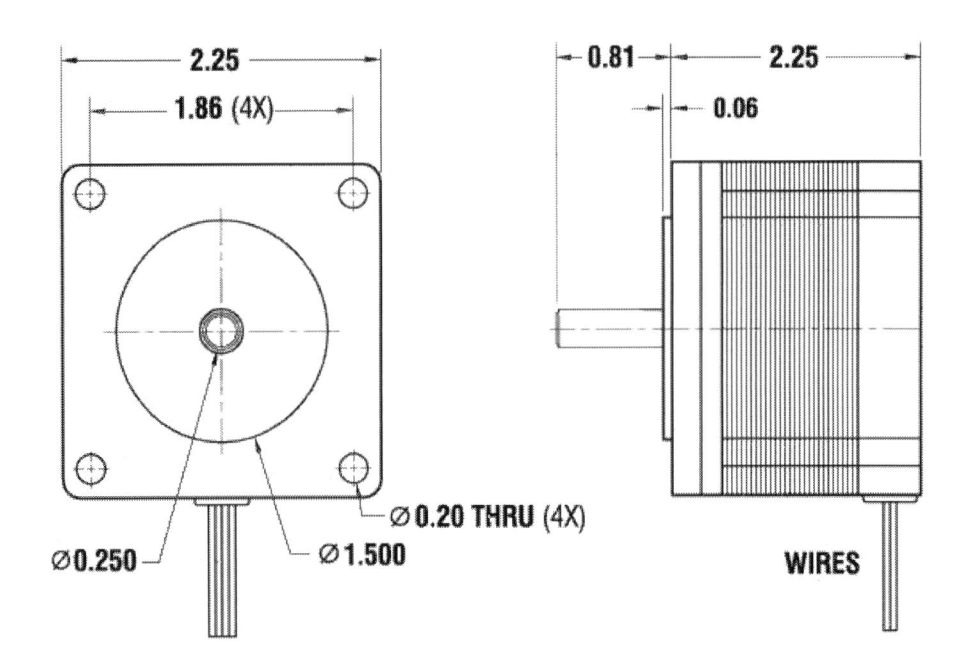

Here is the specification chart for linear rail shaft supports. These support clamps can be used for the shafts as well as for the bearings.

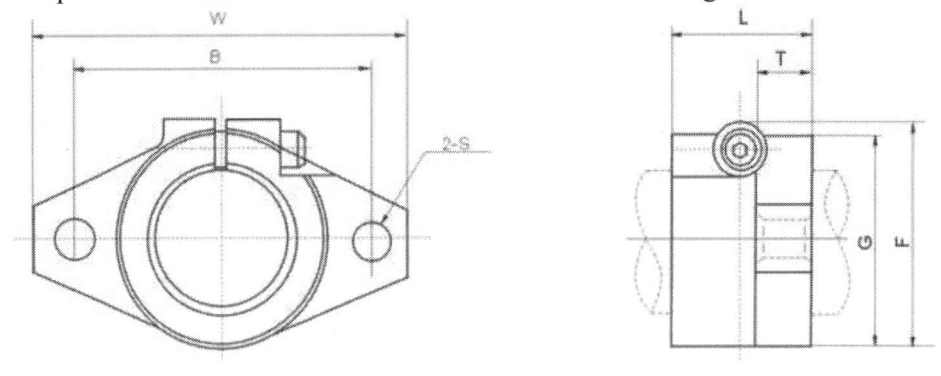

Model	Shaft Dia.	W	L	T	F	G	B	Mtg Bolt
SHF10	10	43	10	5	24	20	32	M5
SHF12	12	47	13	7	28	25	36	M5
SHF13	13	47	13	7	28	25	36	M5
SHF16	16	50	16	8	31	28	40	M5
SHF20	**20**	**60**	**20**	**8**	**37**	**34**	**48**	**M6**
SHF25	25	70	25	10	42	40	56	M6
SHF30	30	80	30	12	50	48	64	M8
SHF35	35	92	35	14	58	50	72	M10
SHF40	40	104	40	16	67	56	80	M10

Another style of shaft support clamp can be used to fasten the 7/8 inch sleeve bearings to the platforms. Here are the specifications for these clamps.

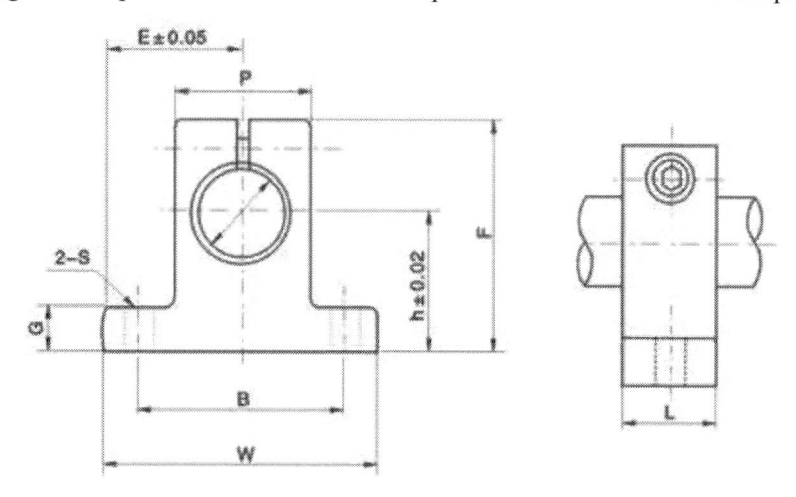

Model	Shaft Dia.	h	E	W	L	F	G	P	B	Mtg. Bolt
SK6	6	20	21	42	14	37.5	6	18	32	M5
SK8/10	8/10	20	21	42	14	37.5	6	18	32	M5
SK12/13	12/13	23	21	42	14	37.5	6	20	32	M5
SK16	**16**	**27**	**24**	**48**	**16**	**44**	**8**	**25**	**38**	**M6**
SK20	20	31	30	60	20	51	10	30	45	M6
SK25	25	35	35	70	24	60	12	38	56	M8
SK30	30	42	42	84	28	70	12	44	64	M10
SK35	35	50	49	98	32	82	15	50	74	M10

On the next two pages there are the mechanical drawings for the front and then the back of the Y axis. The stepper motor mounting holes are not needed on the back piece; instead there are holes for the bearing support. These can be made out of plastic or aluminum. If you are using aluminum, then three inch by one inch angle aluminum would eliminate the need for the additional mounting braces to hold them to the frame or the base plate of the CNC machine. You can also drill holes and tap the plastic to fasten it to the base, but the holes need to be very accurate as you will be drilling very near the edges of the plastic.

The large holes in the drawings are 1/2 inch diameter and the four stepper mounting holes are 8-32 clearance holes. The drawings are around 1/2 of their actual size in scale in order to fit them into this book.

36

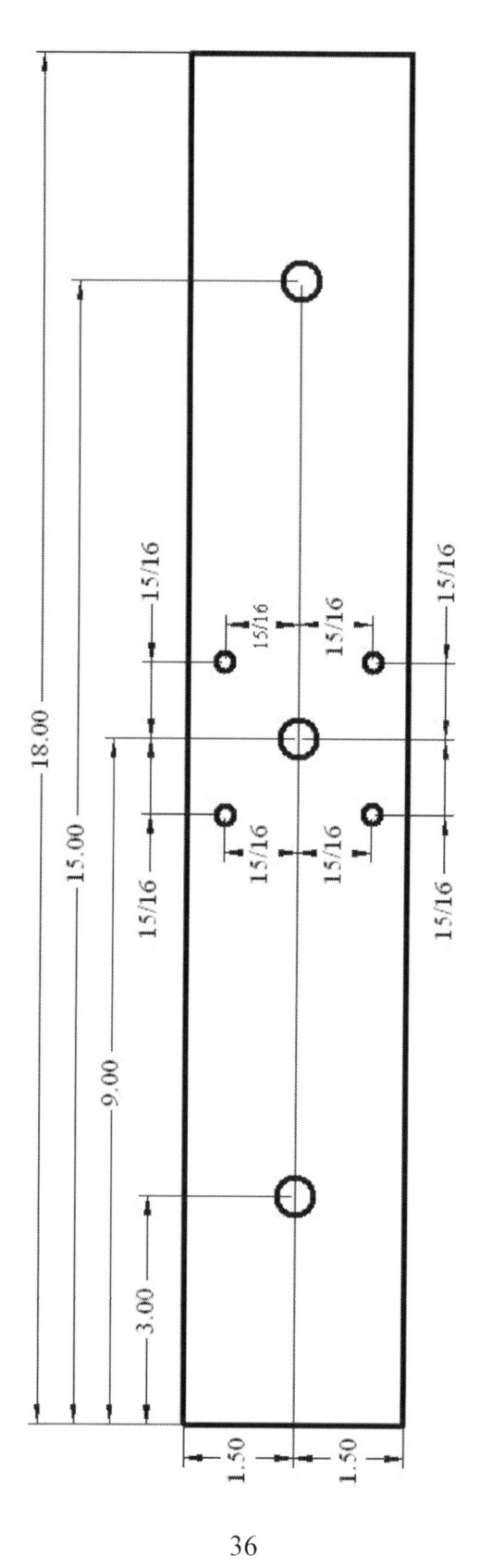

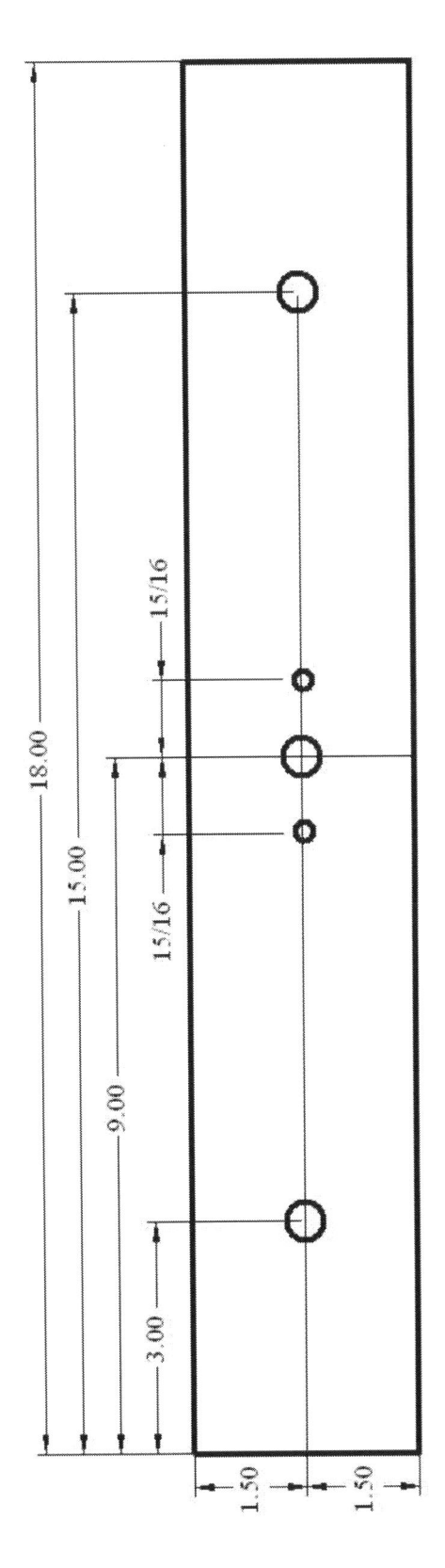

Up next drawing is of the Y platform drawing. Holes should be countersunk.

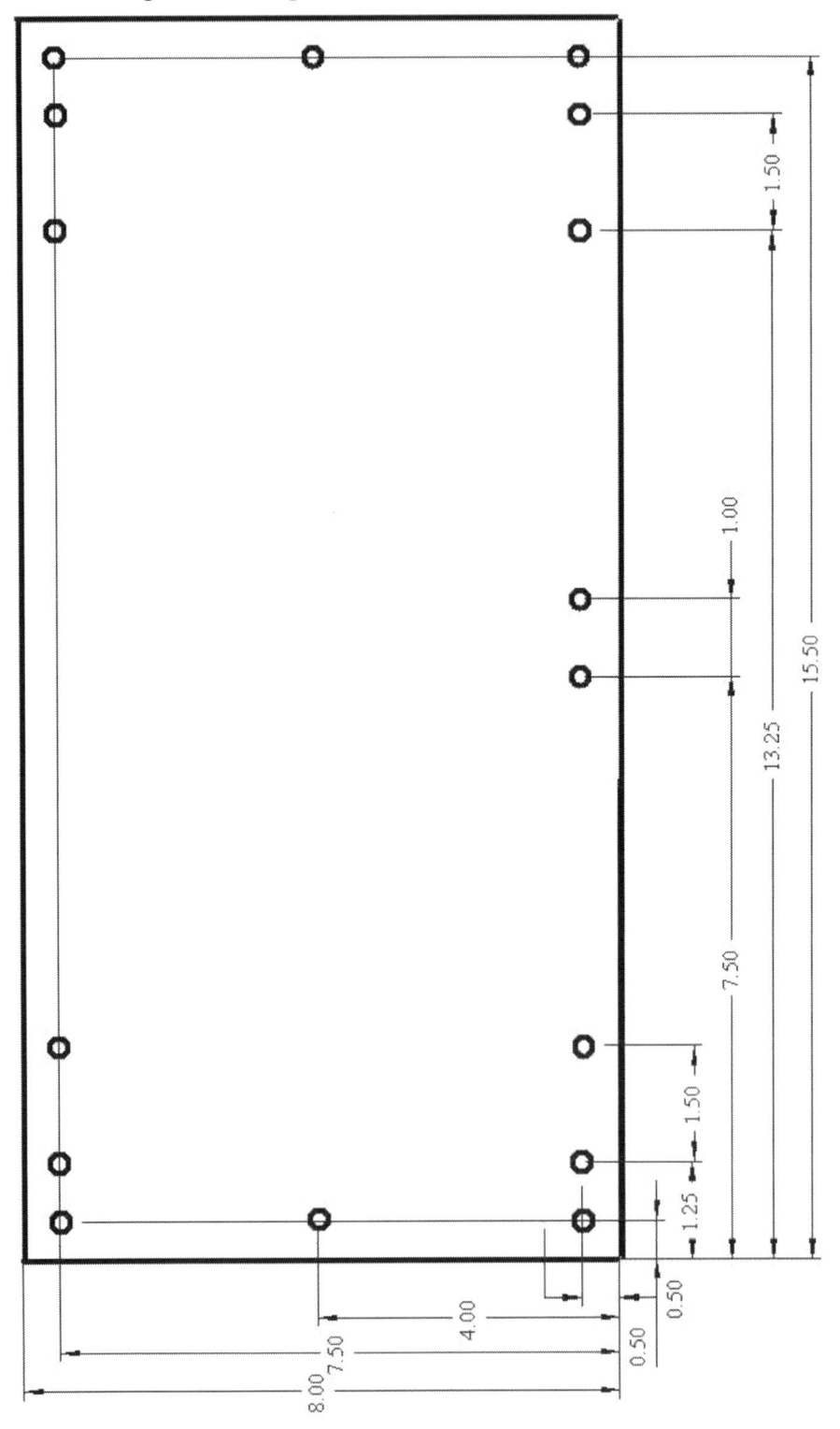

Here is a drawing of the left side of the gantry. The three large holes are 1/2 (or 3/8) inch in diameter depending upon the size of the supporting rods. The stepper motor mounting holes are spaced the same as they were in the Y axis and their size is that of an 8-32 clearance hole or about 5/32 inch.

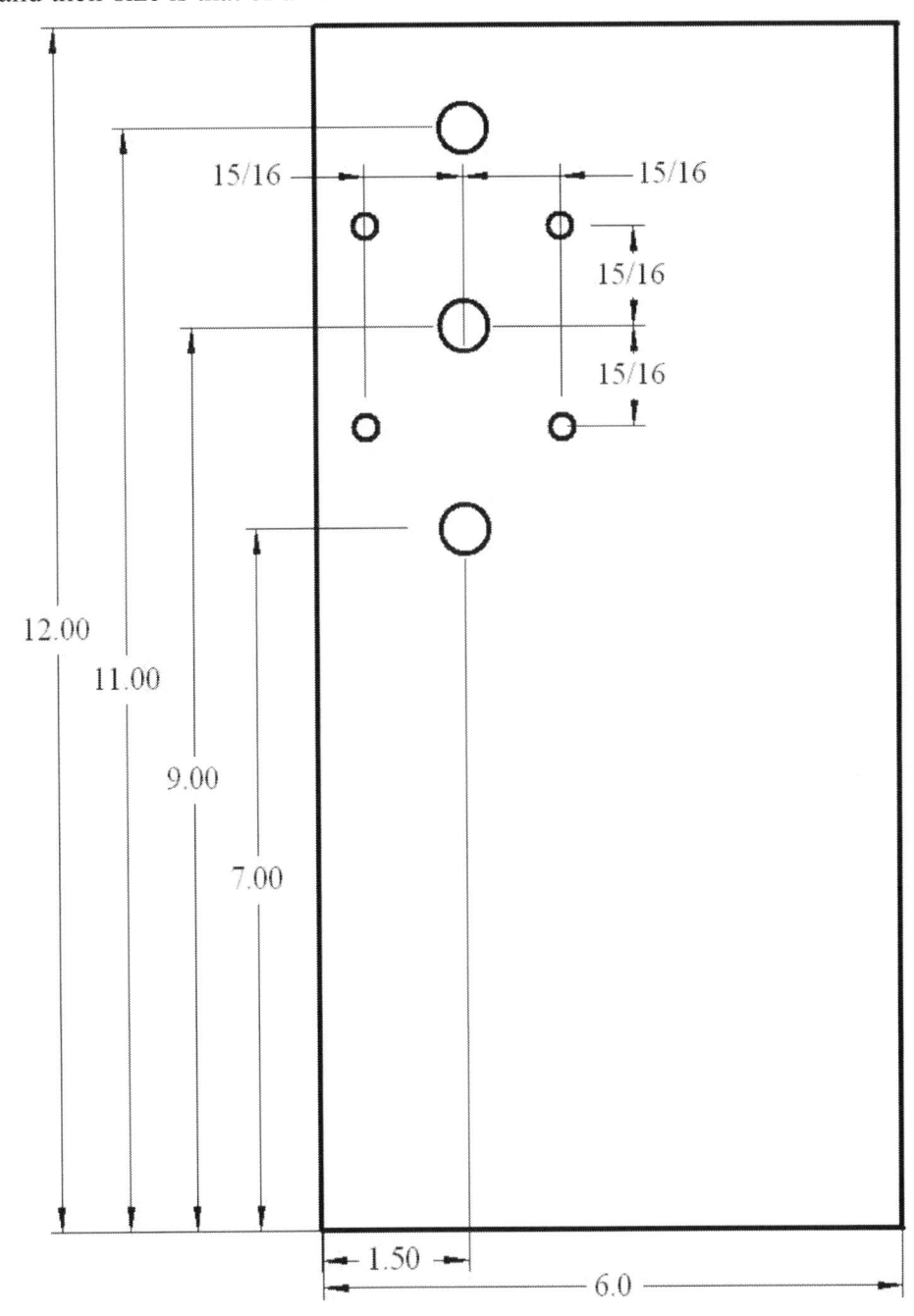

Next up is the drawing of the left side of the gantry. It has bearing support holes instead of stepper mounting holes.

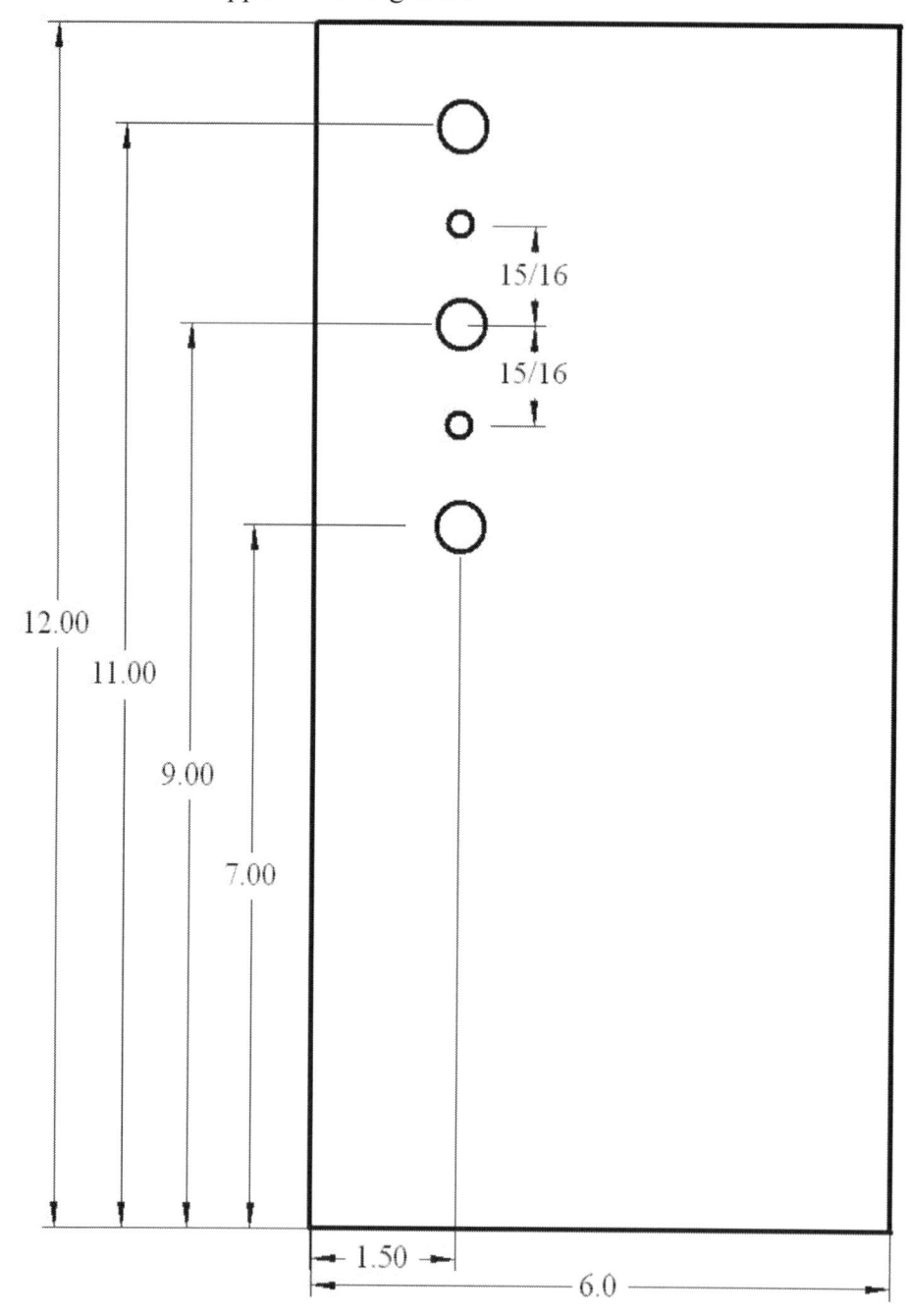

15/16

15/16

12.00

11.00

9.00

7.00

1.50

6.0

The gantry sides can be made out of three inch by 1 inch angled aluminum. In that case it should be spaced three inches from the back of the CNC machine.

The next drawing is of the X axis platform. It is 12 inches tall but 10 inches would have sufficed. In the metal version it is 10 inches tall. The extra 2 inches comes up to the bottom of the stepper motor.

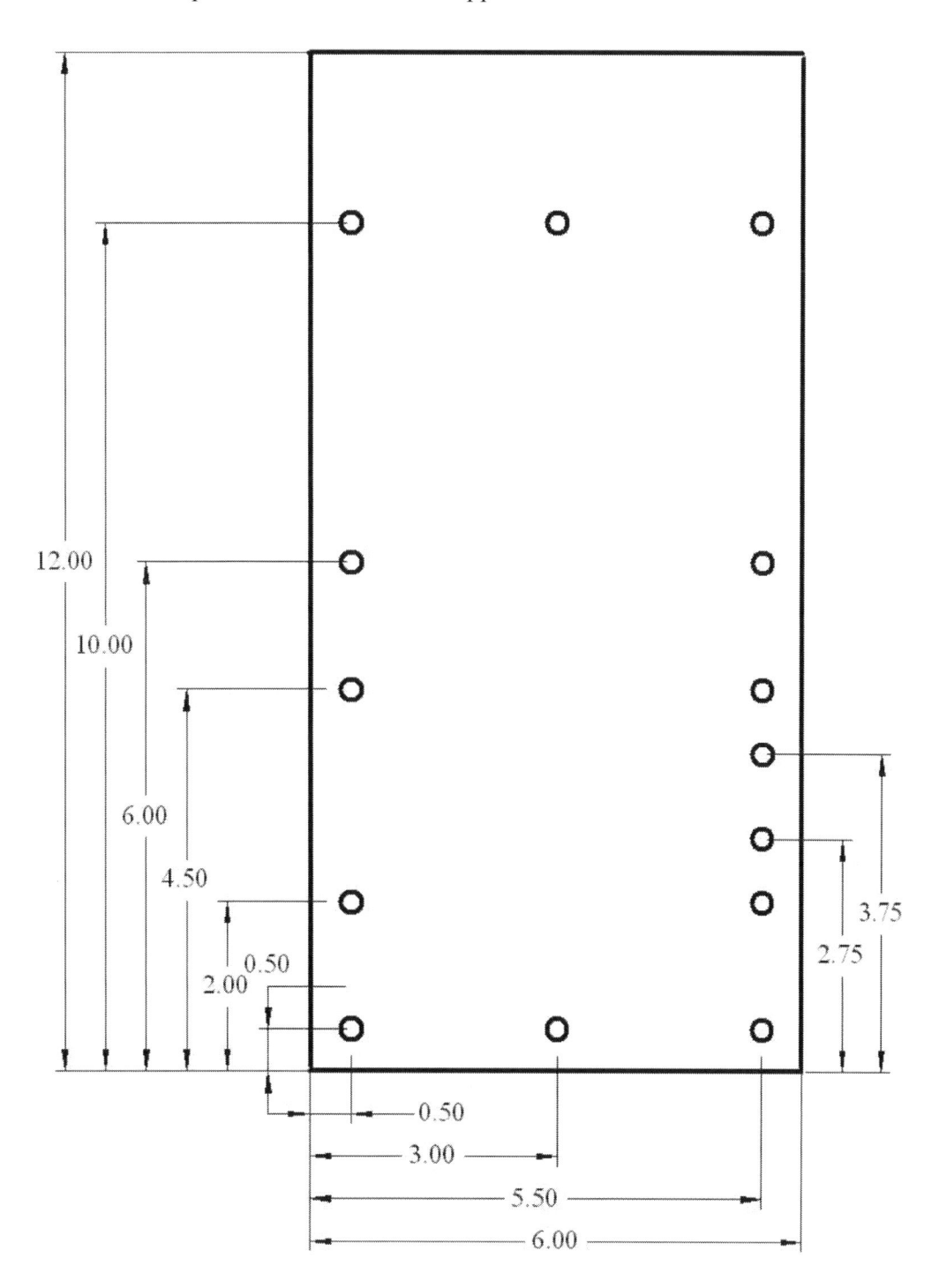

Up next is the drawing of Z axis top and bottom pieces. This is the only assembly that I fastened together by drilling and tapping holes into the plastic. It could be held together with angle brackets just like the X and Y axis were held together.

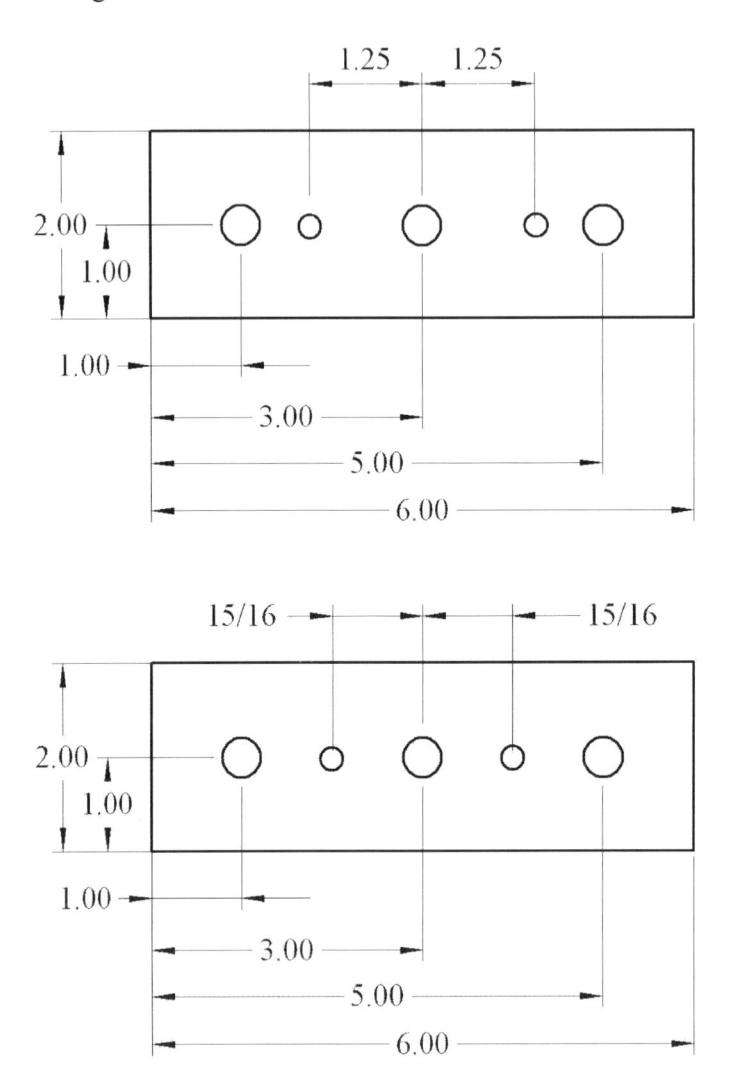

The three larger holes are once again 1/2 inch holes for the guide rods and the leadscrew. The motor on the Z axis is rotated 45 degrees so the two smaller holes on one end hold the motor in place via some 1.5 inch long spacers. The two smaller holes on the other end are for the bearing mounting strap.

The next to last drawing is the Z axis platform.

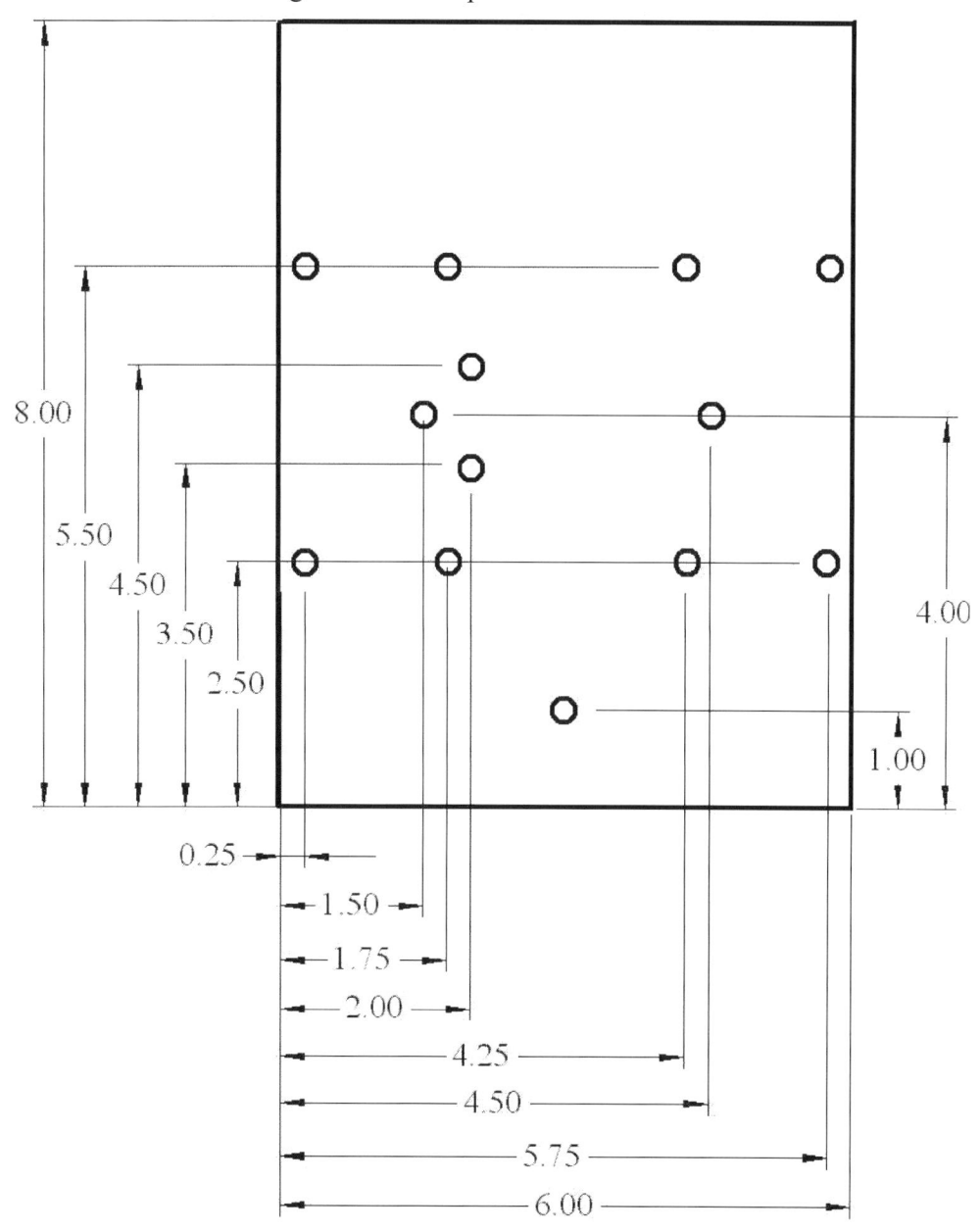

The final drawing is the electronics mounting board. Even if you do not want to mount the electronics here, something is needed to reinforce the gantry so it cannot twist. It has holes for the BOB and four motor controllers. Holes for the mounting brackets are not shown so you can use whatever brackets you want to use to mount it.

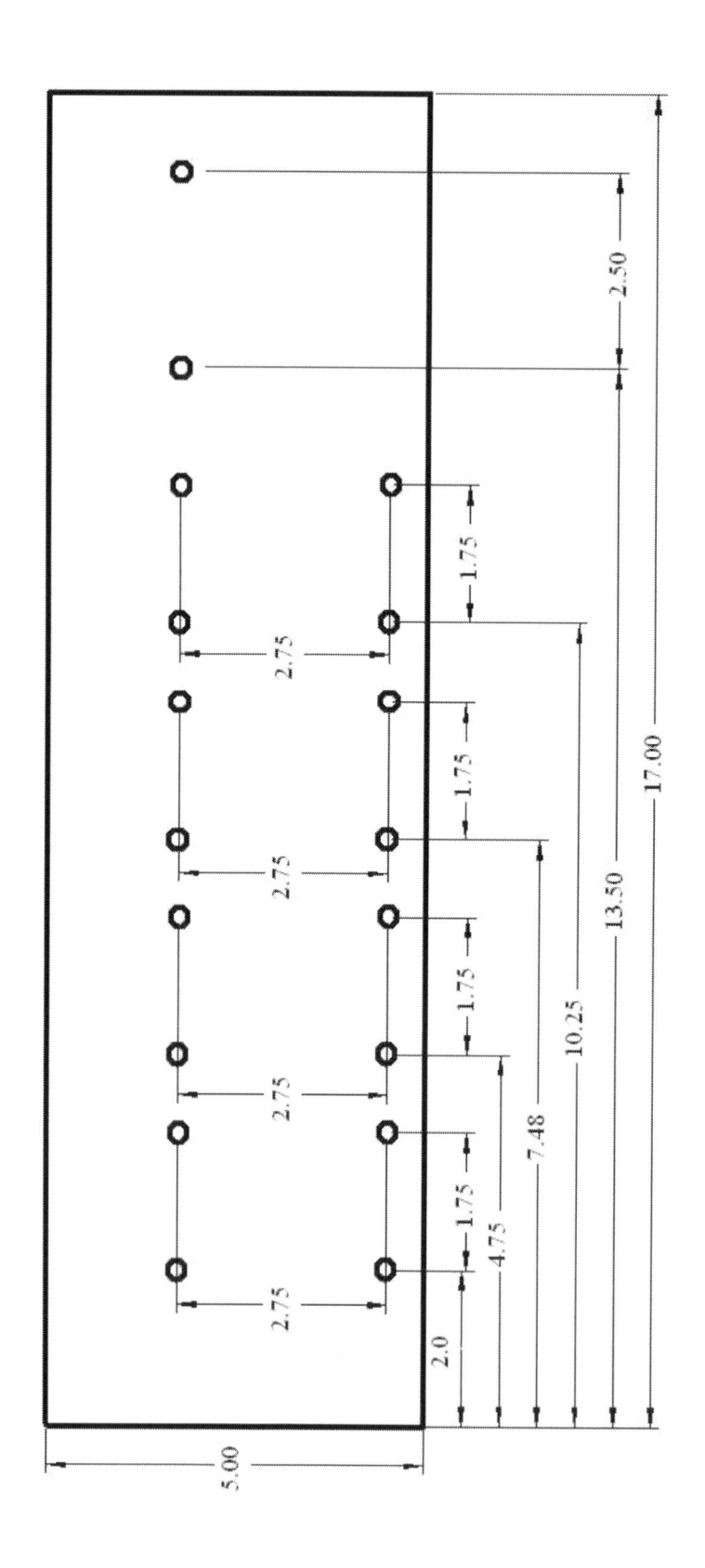

Chapter 4

Mechanical Assembly

The mechanical assembly starts with the Y axis platform. Add the bearings to the bearing holders and then fasten them onto the platform. I used flat head 6-32 screws to keep the platform as smooth as possible. This picture is what the sleeve bearings and 1/2 inch pipe holders look like.

Check the alignment of the rails with the Y axis end pieces and then add the end pieces to the base of the CNC. I did not show the holes for the angle brackets that are used to fasten the ends onto the base in the drawings. That is because you can use any angle bracket that you might want to use. I used one bracket one third of the way down each end of the Y axis end pieces. The brackets were 1.5 inches by 1.5 inches by 1.5 inches in size.

Next you can add the stepper motor, the leadscrew and the end bearing. The stepper motor is mounted with one or two washers spacing it back slightly from the mounting holes as can be seen in the next picture.

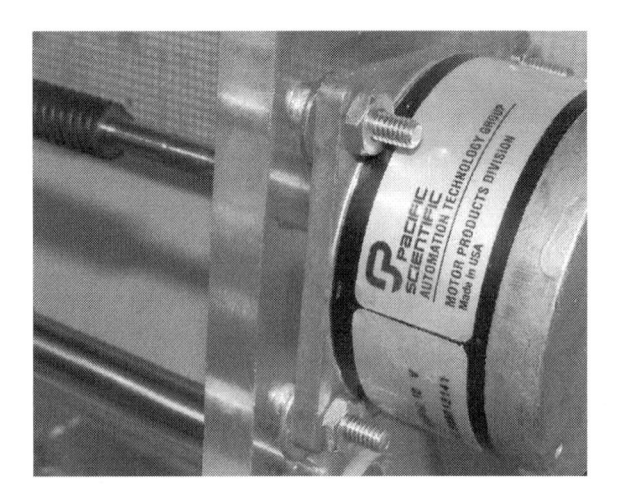

Here is a picture of the leadscrew bearing holder assembly. I put a layer of electrical tape on the outside of the bearing as the holder did not tighten enough to hold it in place.

Up next is a picture showing an alternate way of holding the bearing in place. It is a capacitor mounting strap. This took several layers of electrical tape as the strap was one inch in diameter. This picture also shows collars being used to hold the guide rods in place instead of pipe straps.

Once you have added the leadscrew and stepper motor this picture is what the assembled base of the CNC machine should look like.

The next thing to add is the gantry sides. Once again I did not show the bracket hole positions in the drawings because you can use any bracket that you might want to use. You could use two brackets for each gantry side or use one big bracket for each gantry side. The brackets that I used were three inches long by two inches on one side by two inches on the other side in size. Each bracket used six 8-32 screws. Some of the holes in the brackets were already tapped making the assembly easier. I believe the brackets were meant to be "ears" for adding 19" rack mounting capability to various heavy instruments.

This is a picture of the base with the gantry sides attached.

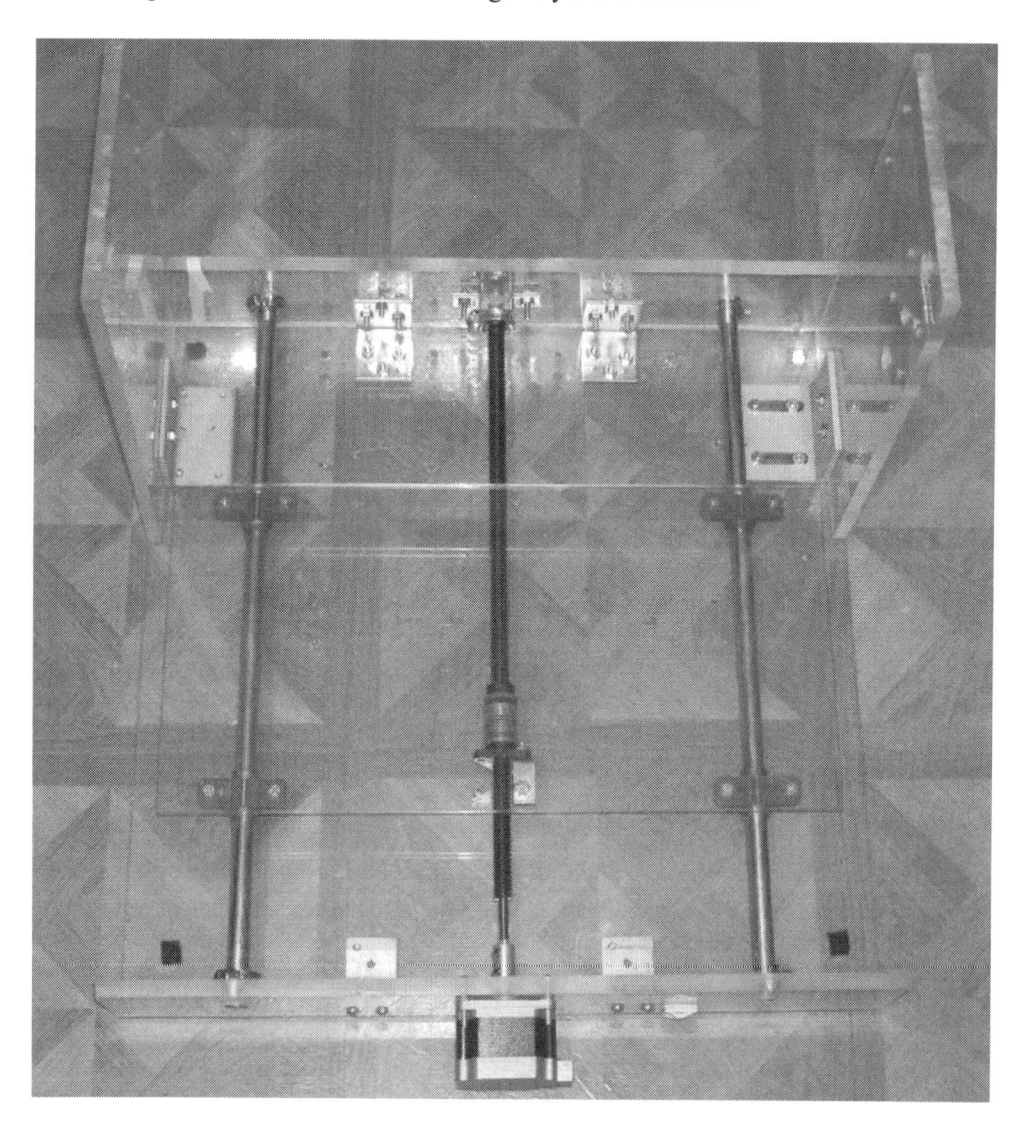

The next thing to be assembled is the X platform and the entire Z axis. This is tricky as there is only a two inch space inside of the Z axis to reach in and put the nuts on the screws.

Start by adding the bearings and supports to the X platform. Then add the bearings and supports to the Z platform. Add the Z support rods and make sure that the Z support rods line up with the Z ends. Then install the Z leadscrew and install the Z end pieces.

This is a front view picture of the Z axis assembly.

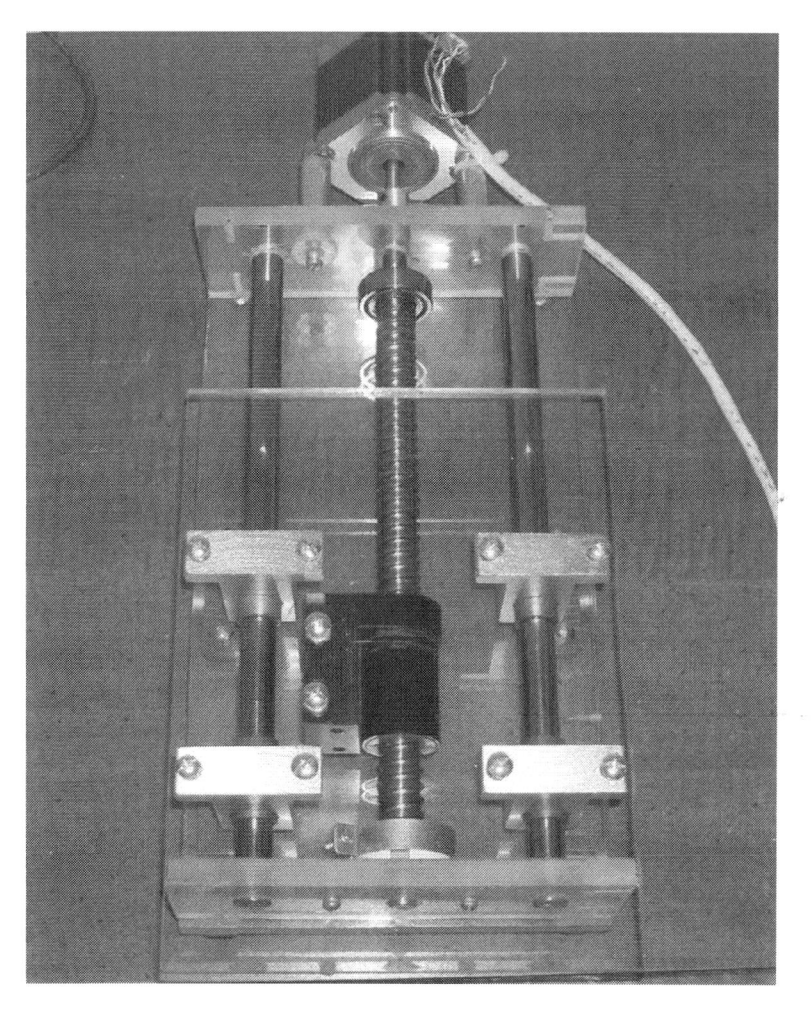

Coming up next is a picture of the side view of the Z axis.

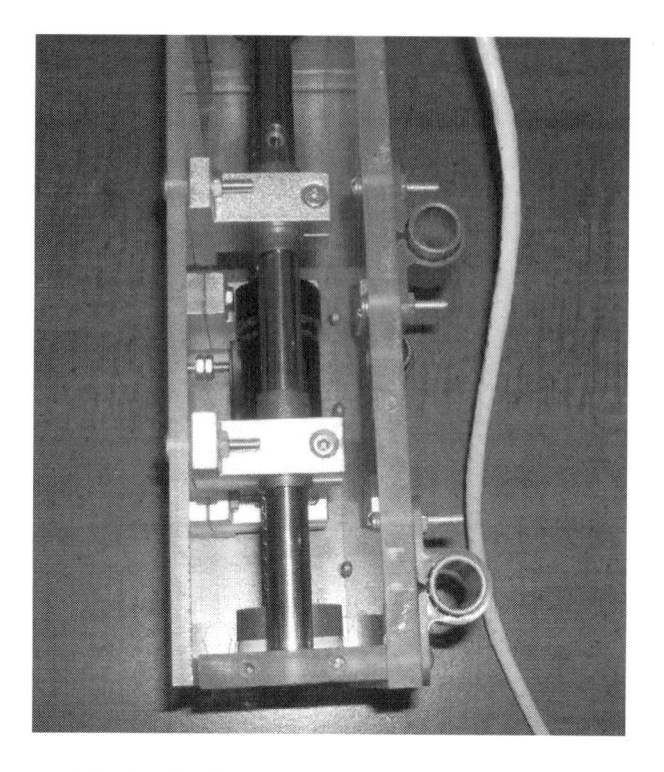

This is a picture of the back view of the Z axis assembly.

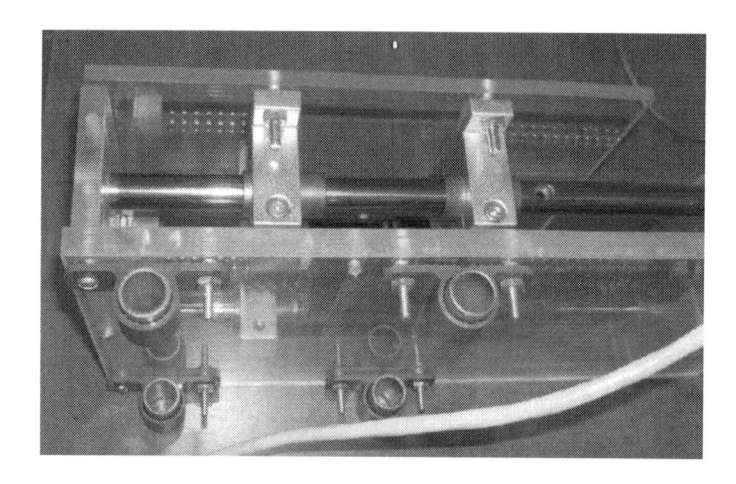

Now you can insert the X axis support rods and check their alignment with the gantry ends. Install the Z axis assembly into the gantry ends with the retaining sleeves. Next install the X axis stepper motor leadscrew and bearing support.

Install the electronics mounting board so that the top edge is flush with the top of the gantry. I spaced mine in an inch from the back edge so the electronics do not hit the wall when it is in storage.

Now you have completed all of the mechanics of a CNC machine! I added the links between the axis platforms and the nut on the leadscrews after the fact, except for the z axis. On the X and Y axis I used small "L" shaped metal brackets. I had to drill and extra hole in the brackets as the existing holes did not come close enough to the bend in the middle of the bracket.

At this point you can also mount a Dremel to the Z axis platform. I used a 1.5 inch pipe holder as it is 1.75 inches in diameter and hence fits the Dremel nicely. For the bottom of the Dremel I used a angle bracket that has been bent at 1/2 an inch to have a "U" shape. The screw to hold this bracket in place has to be a flat head screw so it will not impact the Z axis end stops.

Here is what the mounted Dremel looks like.

The bits that come with the Dremel do not work well for a milling operation. Here is a picture of the Dremel "High Speed Cutter" bits that I purchased. They closely resemble the bit used in LinuxCNC but there is a problem with them. Because they go down to a point the teeth get very small. As a result they do not remove much material per revolution and they will need to use a slower feed rate.

Coming up in the next chapter is the electronics assembly.

Chapter 5

The Electronics

The electronics design changed a lot since I started building this CNC machine. When I started out there was an Arduino and two L298 driver boards for the electronics. There was a problem or two with that setup. For one thing in order to run the three steppers with four wires each required 12 pins of the Arduino. That left six pins for everything else. For another thing, every CNC design that I looked at used stepper controllers that responded to "step" and "direction" commands.

Here is a picture of that first setup. The variable resistor controlled the speed of the stepper motors.

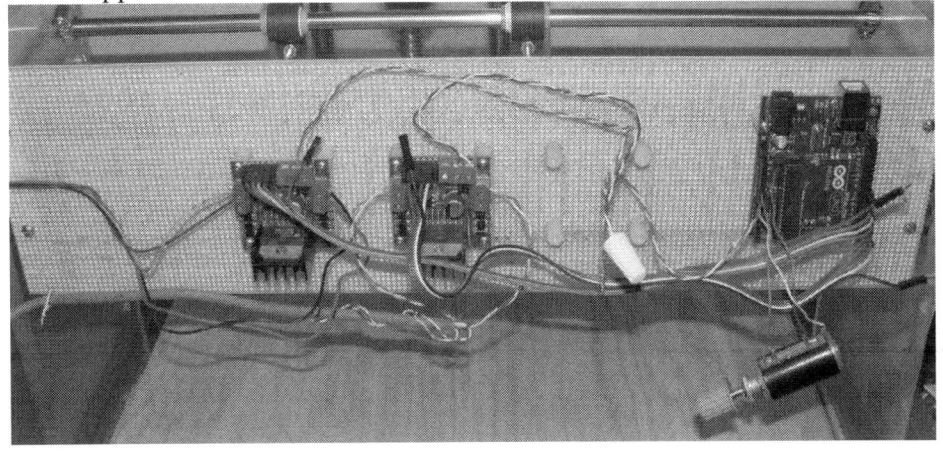

To get around those first limitations I ordered some L297 chips and built an interface board to go between the Arduino and the L298 drivers. I was unable to get that board to do much of anything.

Next I built a L297 and TIP120 driver board. I even included optocouplers on the inputs. When that board came to life it turned on all four phases of the stepper motor and overloaded the power supply. One of the problems was a bad connection to one of the IC pins. This driver requires a five or six wire stepper motors with the centers of the windings going to +12 Volts. The

TIP120's could be replaced with a L298 and then it would work with four wire stepper motors.

Here is the schematic of the L297 Driver board. The resistors in the bottom right corner are .5 ohm at one watt.

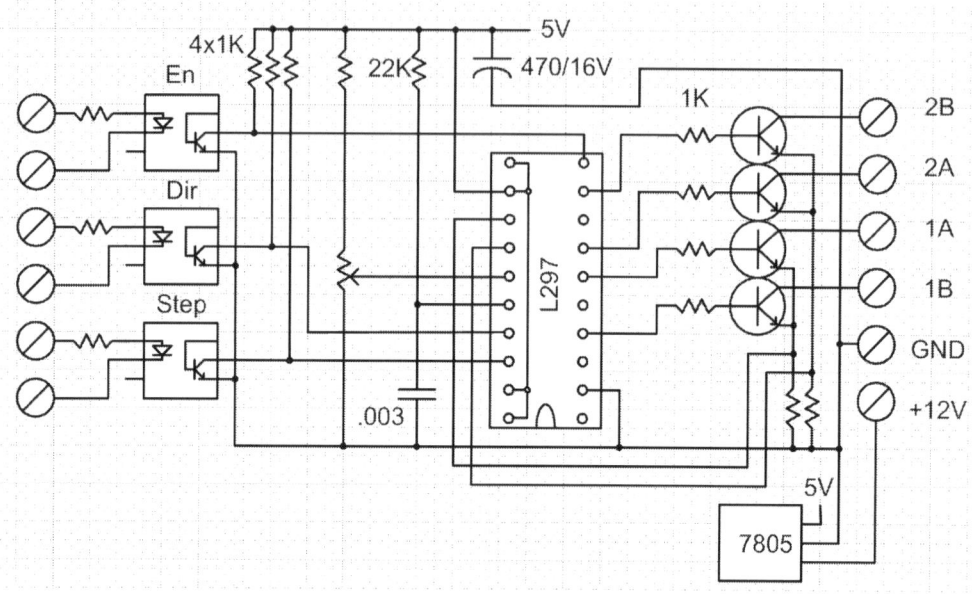

Here is a picture of the L297 driver board. I never added the enable optocoupler.

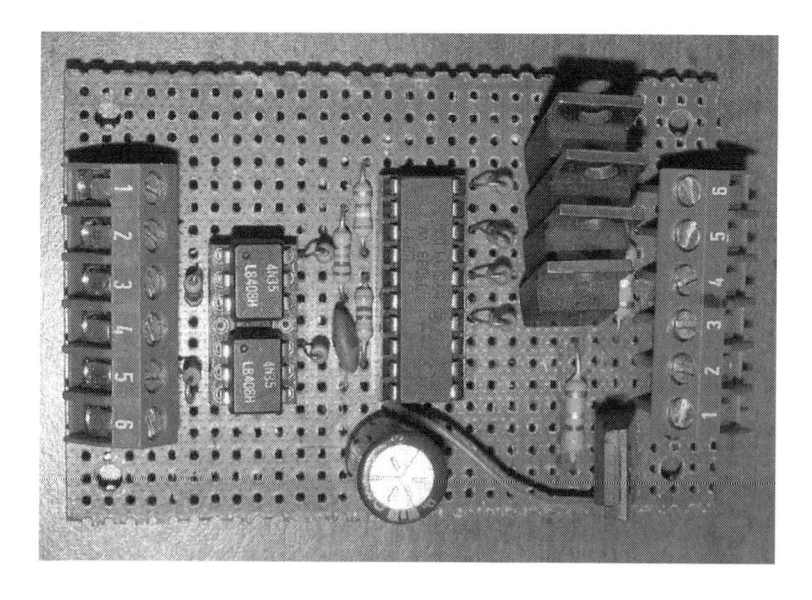

I also build a driver using the "Step Stick" or "Pololu" or "A4988" drivers. They require very few components to interface them to the CNC. This board worked fine from the first try. However, it lacked power to raise the Z axis causing some skips. Here is a picture of that board.

Here is that schematic of the Step Stick interface. I left out a second 470/25V filter capacitor that is on the 5V power supply. The resistors are 1K ohm.

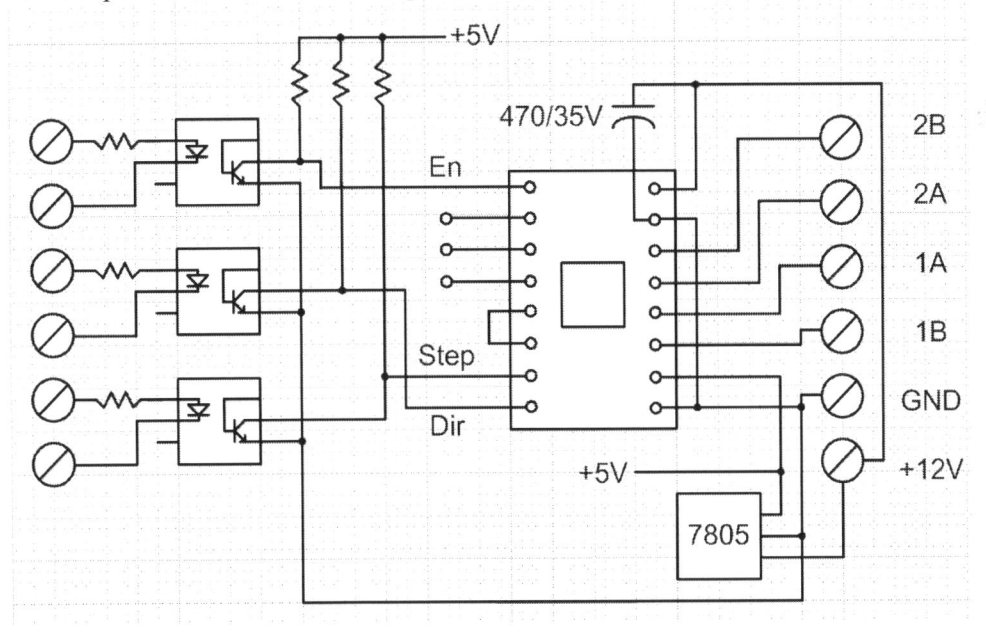

At some point I ordered a couple of the TB6560 driver boards. They featured optocoupler input protection, a controller and a driver all on one small board. The first tests with this new controller revealed that it was a winner! They worked great right from the start. Initially I set them up with 1.6 amps of driving power. That is done by turning off all switches but SW1, the top one.

Across the top of the driver board, the left four terminals go to the stepper motor. The next connection is ground and then the last connection is the 12 or 24 volt power pin.

Across the bottom there is the enable (EN), the direction (CW) and the step (CLK) inputs. You can connect the positives of those three inputs to 5V on the break out board.

The next picture is what the TB6560 motor driver board looks like.

Here are some charts that show some of the TB6560 switch settings. In this first chart SW1 should be "on" and all the others "off" for 1.6 amps for a typical NEMA 23 sized stepper motor.

Running Current														
(A)	0.3	0.5	0.8	1	1.1	1.2	1.4	1.5	1.6	1.9	2	2.2	2.6	3
SW1	OFF	OFF	OFF	OFF	OFF	ON	OFF	ON	ON	ON	ON	ON	ON	ON
SW2	OFF	OFF	ON	ON	ON	OFF	ON	OFF	OFF	ON	OFF	ON	ON	ON
SW3	ON	ON	OFF	OFF	ON	OFF	ON	ON	OFF	OFF	ON	ON	OFF	ON
S1	ON	OFF	ON	OFF	ON	ON	OFF	ON	OFF	ON	OFF	ON	OFF	OFF

The stop current has to be enough to hold the motor in place when it is not moving.

Stop Current	
	S2
20%	ON
50%	OFF

The "Excitation mode" should be set to "Half Step" or S3 "on" because most software expects half stepping to be enabled.

Excitation Mode		
	S3	S4
Full step	OFF	OFF
Half Step	ON	OFF
8 subdivision	ON	ON
16 subdivision	OFF	ON

Decay Setting		
	S5	S6
0%	OFF	OFF
25%	ON	OFF
50%	OFF	ON
100%	ON	ON

Here is an over simplified schematic of the TB6560 driver. They did not follow this design as there is another IC on the board, perhaps it is a 74LS14 Schmidt trigger?

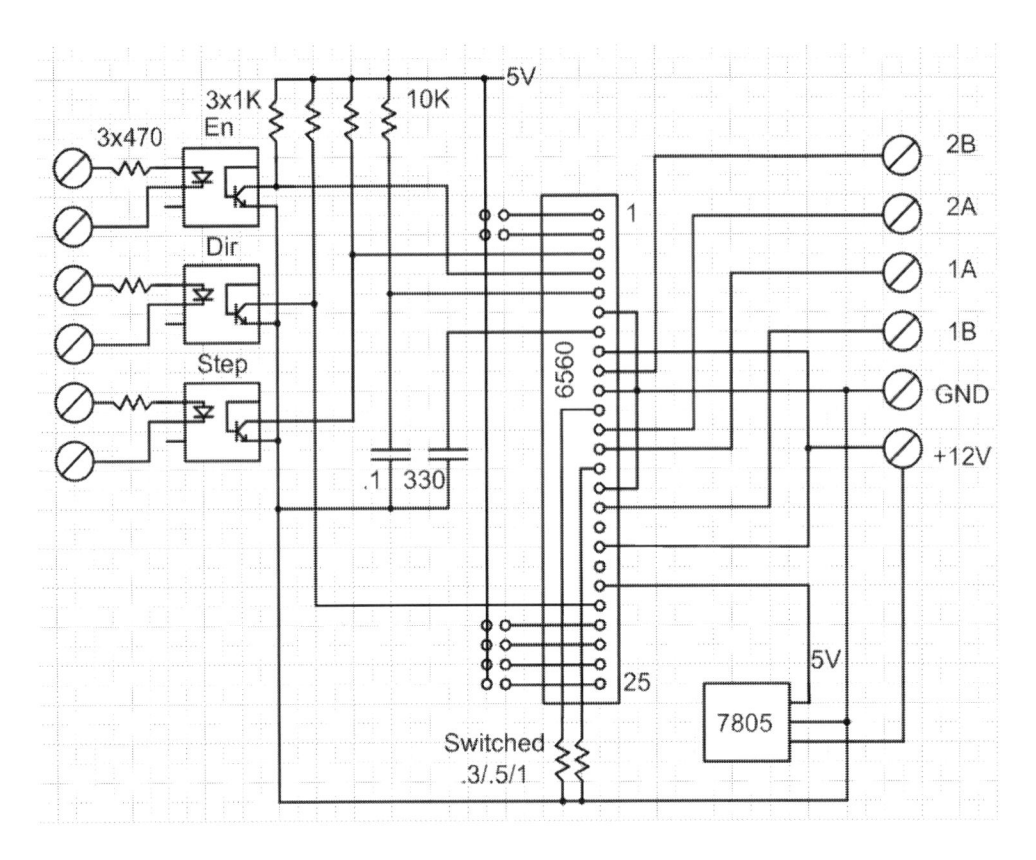

Once I started using LinuxCNC I discovered that there is a problem with using the TB6560 driver boards. They only work on the slowest two settings! In fact I found this chart (shown below) that verified that the TB6560 is one of the slowest stepper motor controllers out there. I greatly reduced size of the chart to only list one or two steppers from each manufacturer.

Also note from this chart that the "Pololu" stepper controllers are much faster, hence I want to eventually try one out. Their big limitation is that they can only handle 1 amp or 1.5 amps with a heat sink attached.

Manufacturer	Model	Step Time	Step Space	Direction Hold	Direction Setup
Chinese Blue Boards	TB6560 Stepper Controller	150000	150000	150000	150000
Gecko	201	500	4000	20000	1000
Granite Devices	VSD-E/XE Evolution	125	125	125	125
JVL	SMD41	500	500	2500	800
Linistepper Open Source	RULMS1	30000	100000	4000	2500
*Motion Control	MSD542	>1500	2000	2000	4000
Parker	OEM750	200	300	0	1000
ST	L297	?	500	4000	2000
Xylotex	XS-3525/8S-3	2000	1000	200	200000
Lin Engineering	Silverpak 17D/DE	20000	20000	200	1000
Hobbycnc	Pro Chopper Board	2000	2000	2000	200
*Routout	2.5amp Stepper Driver	200	1000	1000	200
*Intelligent Motion System	IM483	1000	1000	1000	200
Keling	4030	5000	5000	20000	2000
Sherline	8760	1000	6000	24000	?
Burkhard Lewetz	Step3S	6000	15000	?	1000
Parker Compumotor	Zeta 4	200	200	?(200)	20000
www.cncdrive.com	Dugong	1000	2500	1000	10000
www.cncdrive.com	DG2S-08020	1000	2500	1000	24000
Wantai Motors	DQ542MA	5050	5050	500	5000
Leadshine USA	Digital DM422 40V 2.2A	7500	7500	20000	?(200)
Leadshine USA	Brushed servo DCS303	2500	2500	10000	1000
Pololu	A4988 Stepper Motor	1000	1000	200	1000
Pololu	DRV8825 Stepper Motor	1900	1900	650	500
cnc4you	CW5045	2000	8000	5000	5000

Here is a picture the final setup of the electronics. I have rotated the printer port interface or BOB board by 90 degrees to simplify the wiring. That unfortunately makes the printer cable go straight up into the air.

Another design problem is that there are dozens of pin arrangements for interfacing to an Arduino. However, there are standard pinouts available for interfacing a CNC to a parallel port of a PC. Also, testing the machine out is a lot easier when it is interfaced to a PC.

Once the Arduino was gone I needed to come up with a "Break Out Board". I call it Bob's BOB. Basically, the first 18 pins are bought out to screw terminals. There is also a power distribution circuit that was added to the BOB. This picture was taken before the pull-up resistors were added.

On the next page is the schematic diagram of my break out board or "BOB". I had to add 470 (or 1K) ohm pull up resistors on all outputs and I added 10K pull up resistors on all inputs (the limit/home switches). I found out later that the input pull up resistors can also be 470 ohms. The pull up resistors reduced the occasional "miss" on the stepper motors and made them able to work much faster.

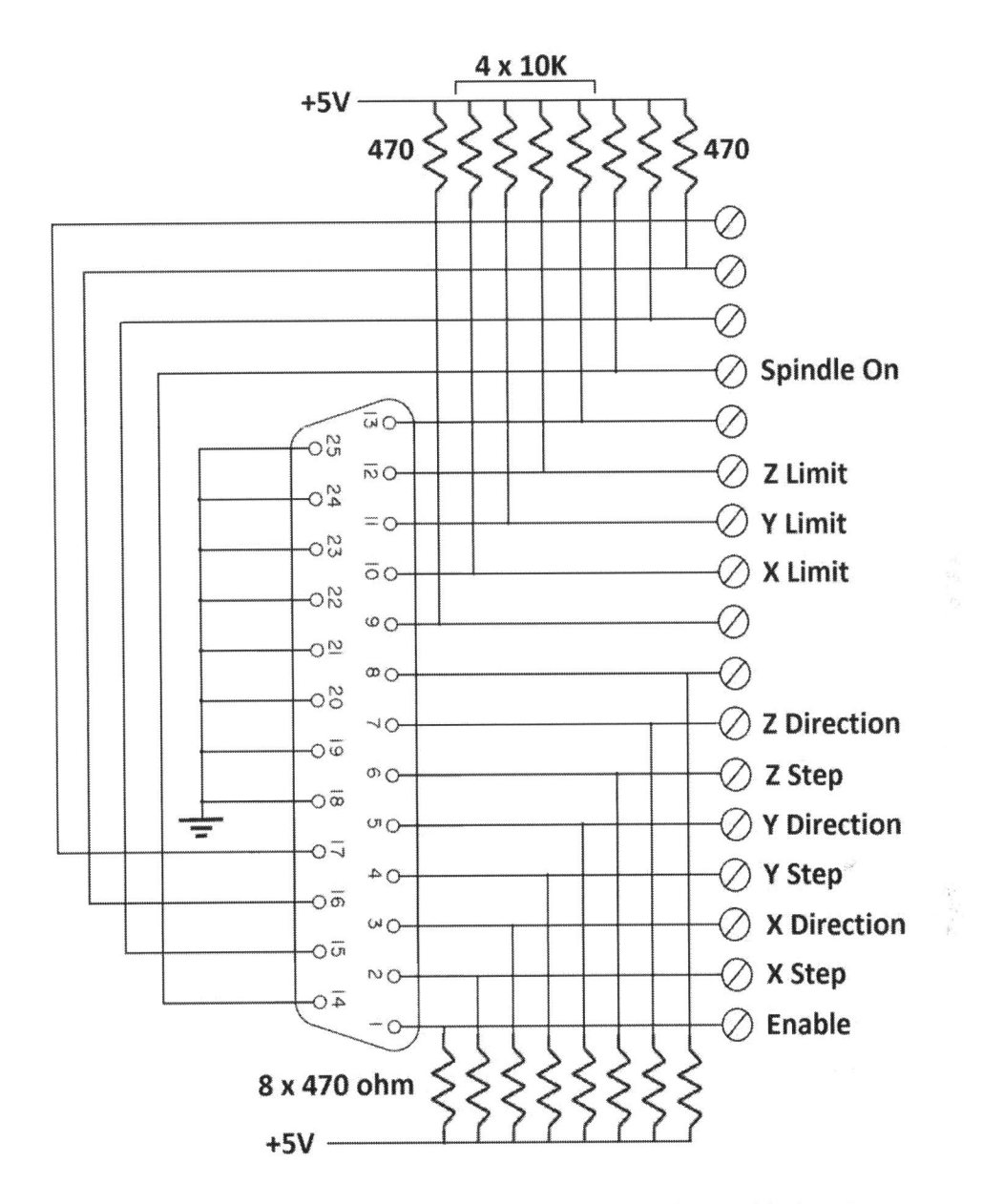

Here is the schematic of the power distribution circuit that I added to the break out board. I forgot to add a very necessary four or five amp power fuse to prevent a possible meltdown. Power input is either 12 volts DC or 24 volts DC at about five amps.

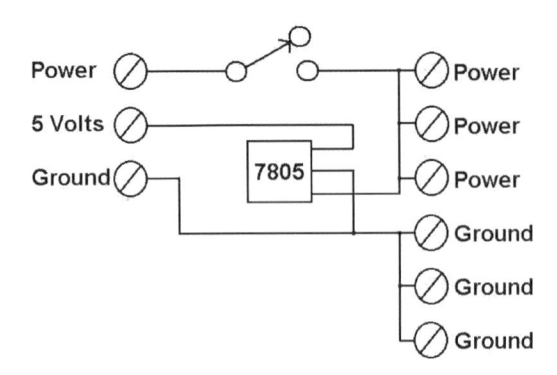

Up next is a complete wiring diagram for the CNC. This schematic shows how everything interconnects so it necessarily takes up an entire page. It is complete except for the home or limit switches. The home switches go to pins 10, 11, and 12 and to ground (pin 18) at the other end.

A last-minute change is not in the schematic bit I added two 470 ohm resistors in parallel between the bob board pin one and the enable inputs of the stepper motor controllers. They were added because without them the enable "High" of about 3.5 volts was not enough to turn off the optocouplers in the stepper motor controllers. Without the resistors, the stepper motors remain off line unless you disconnect the enable inputs. Any resistor from 220 ohms to 470 ohms should do. A diode might do as well.

The schematic is also missing the power input connections and five volt output connection that are on the break out board. They are included in the break out board schematic above.

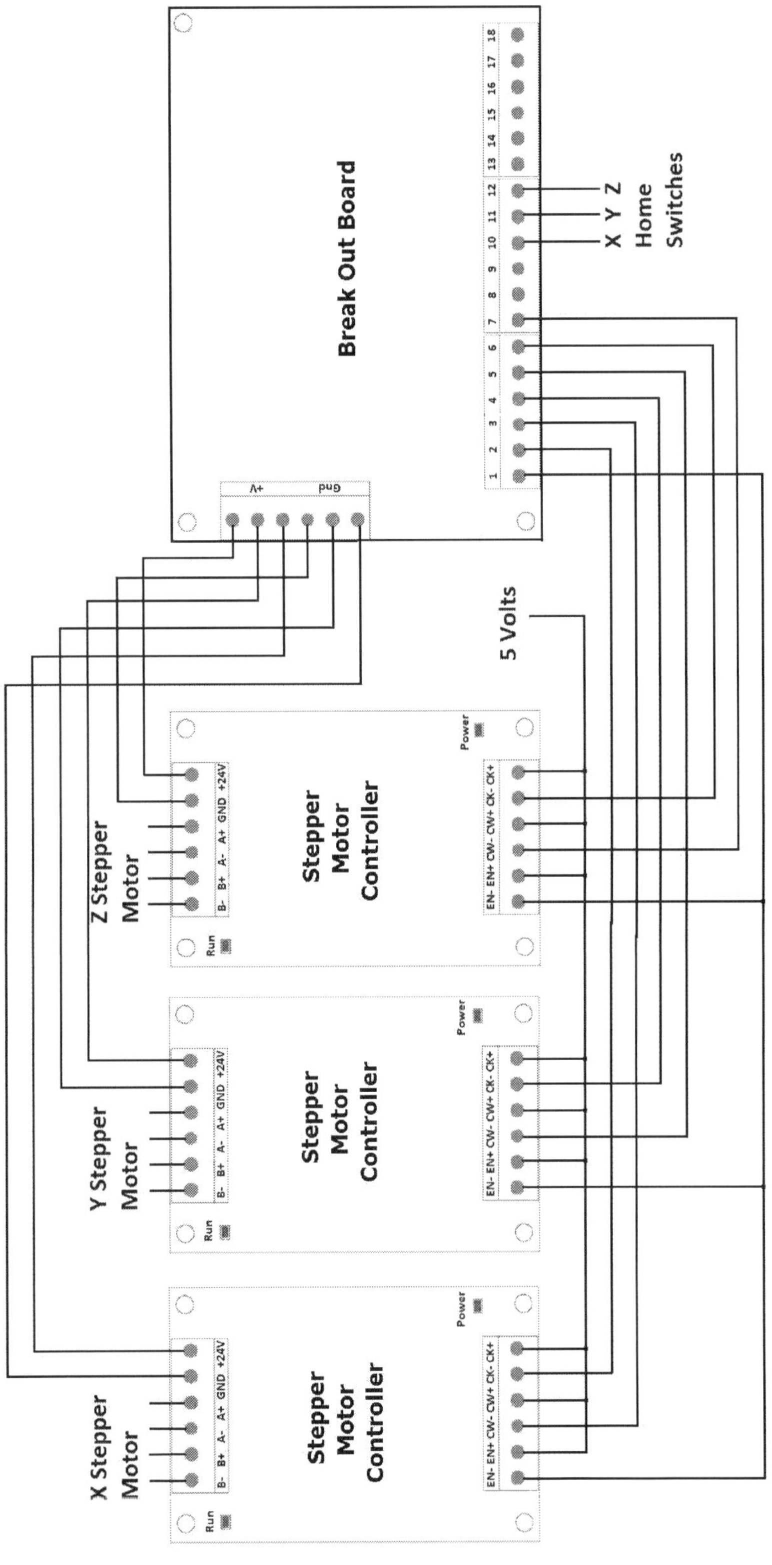
Break Out Board
+V
Gnd
X Y Z
Home
Switches
5 Volts
Z Stepper
Motor
Y Stepper
Motor
X Stepper
Motor
Stepper
Motor
Controller
Stepper
Motor
Controller
Stepper
Motor
Controller
Power
Run
B- B+ A- A+ GND +24V
EN- EN+ CW- CW+ CK- CK+

Home/Limit Switches.

The CNC can run without limit or home switches but who wants to calibrate it every time it is turned on? I used magnetic switches for the X and Y axis since they can work from over half an inch away. Magnetic switches do not work when cutting magnetic metals as they will get covered with small pieces of metal.

The Linux CNC program is supposed to allow limit switches to double as home switches. The combination of limit and home switches is hard to get working. It is supposed to move the CNC axis to the limit switch and then it measures out the distance to the home position. It is easier to have the switches just be "home" only switches.

First we need to get our X, Y, and Z axis orientation correct.

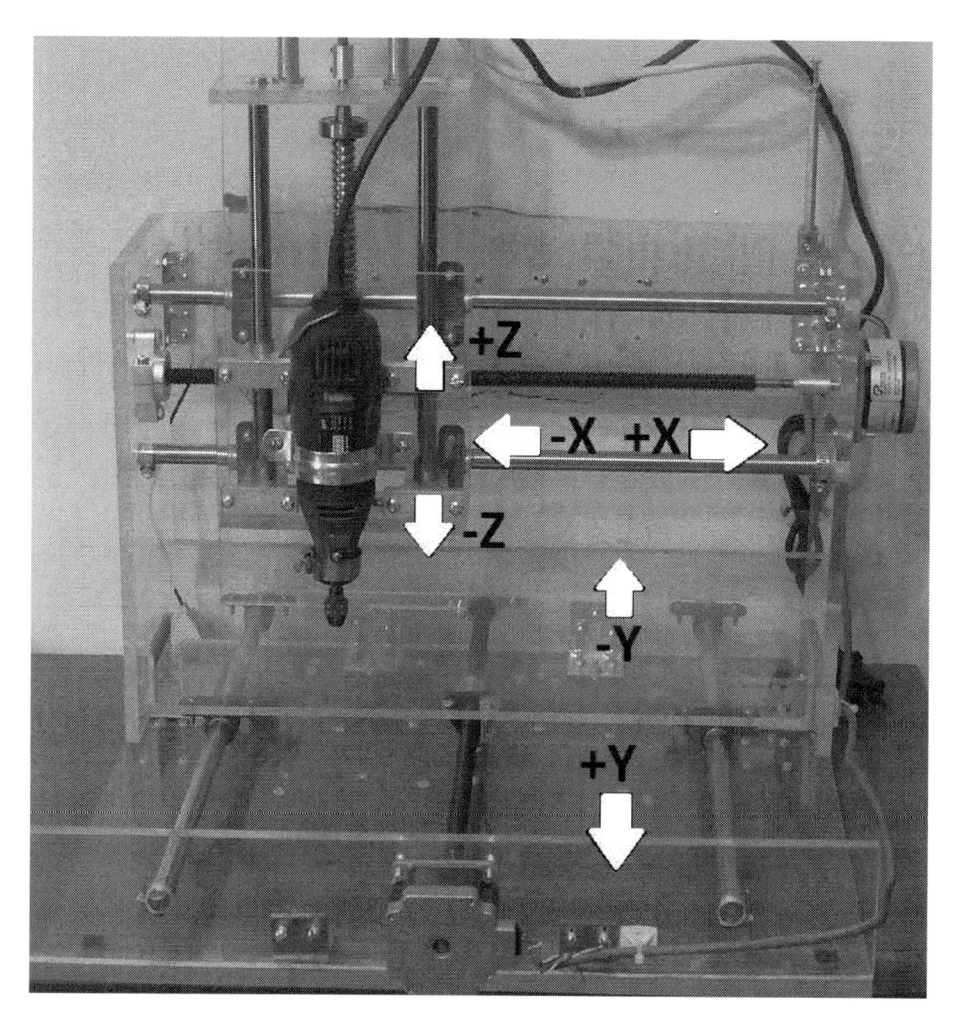

You will note in the above picture that the "Y" axis is marked backwards. The drawback with a fixed gantry is that any axis that is not on the gantry is reversed. When the Y axis moves away from you, the point being milled moves towards the front edge. So the Y axis is backwards from what you might expect.

This is the X axis home switch mounting on the left side of the gantry.

Here is the Y axis home switch mounted behind the machine. The "0" position of the Y axis is located at the back of the CNC machine.

Last off all there is the Z axis home or limit switch. I used a mechanical switch as the Z axis is usually calibrated at .1 inch from the top of the material that is being worked on. I have never used this switch.

Chapter 6

CNC Pendant

To test the CNC machine out without using a computer I created my own "Pendant". The pendant was made out of an old Zenith TV remote control. The critical parts that are needed are some three position toggle switches. The center position of these switches is "off". I had several switches lying around that are also "momentary contact" as in the lever always returns to the center.

This is what the completed CNC pendant looks like. The cable used to connect to the CNC is a stranded or flexible four pair network cable.

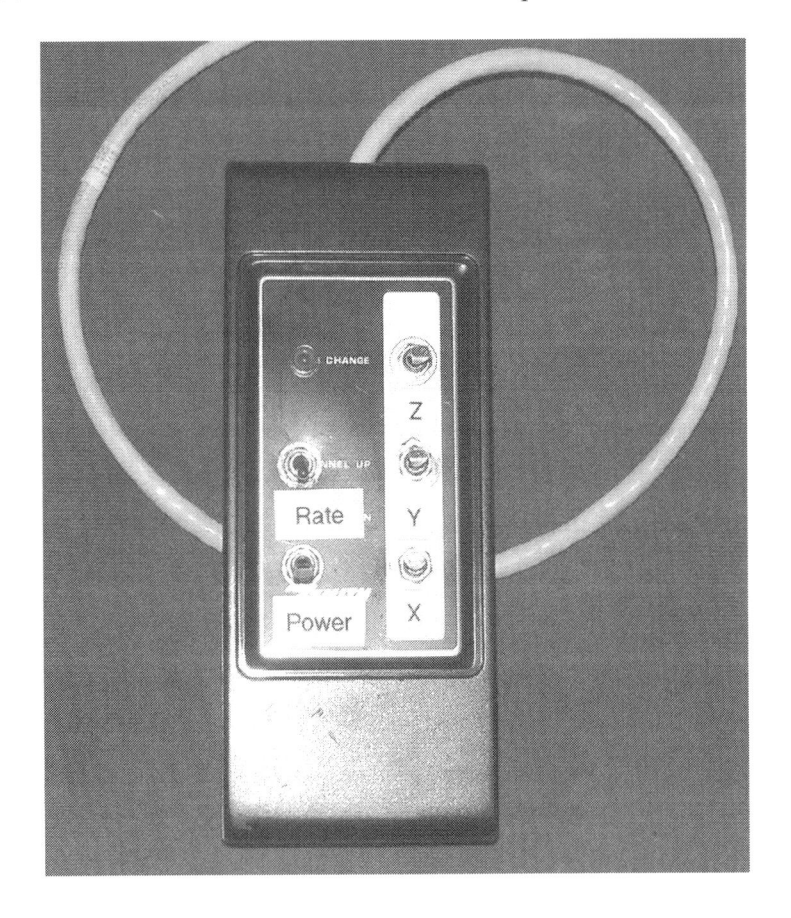

The next picture is what the insides of the pendant look like. The wiring is relatively simple and straightforward. The power source is a used 9 volt battery that supplied about 6 or 7 volts. Later on I added a LM7805 5 volt regulator IC. Everything in this CNC design is compatible with 9 volts but some other BOB interfaces might be damaged by that high of a voltage.

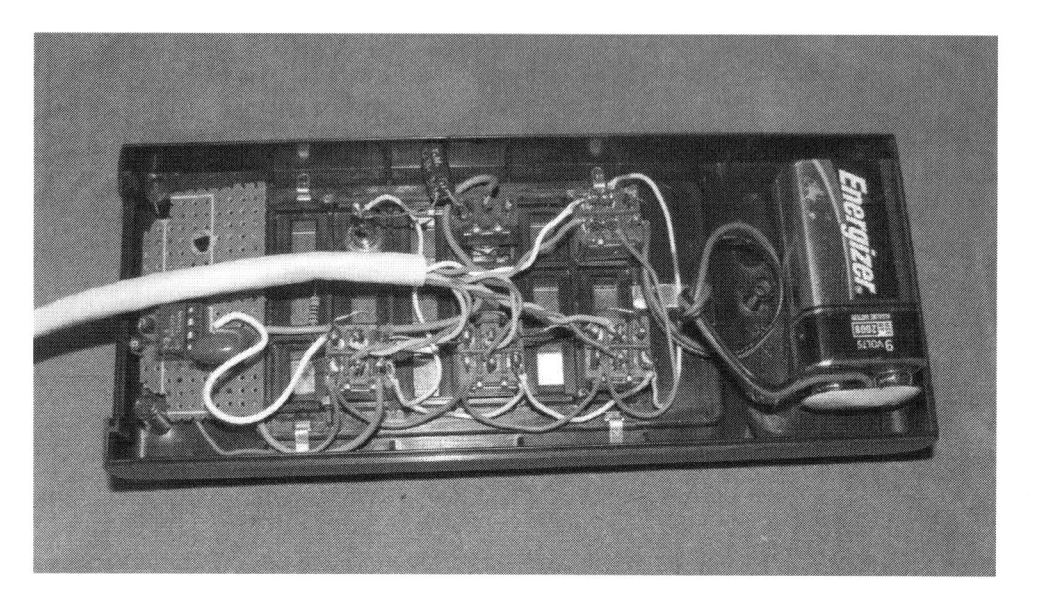

Here is the schematic diagram of the X, Y and Z switches portion of the pendant. Basically, each switch selects a direction and connects the clock signal to the clock input.

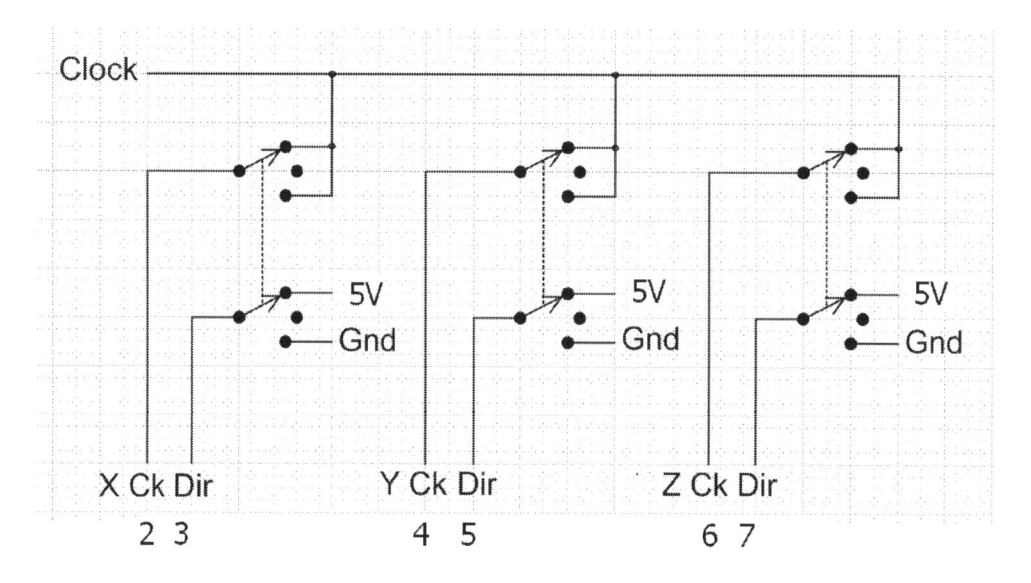

68

This is the schematic diagram of the power supply and the 555 timer for the pendant. The 555 makes it possible to rapidly move the CNC axis around. For this CNC 400 to 500 HZ is about as fast as it can go.

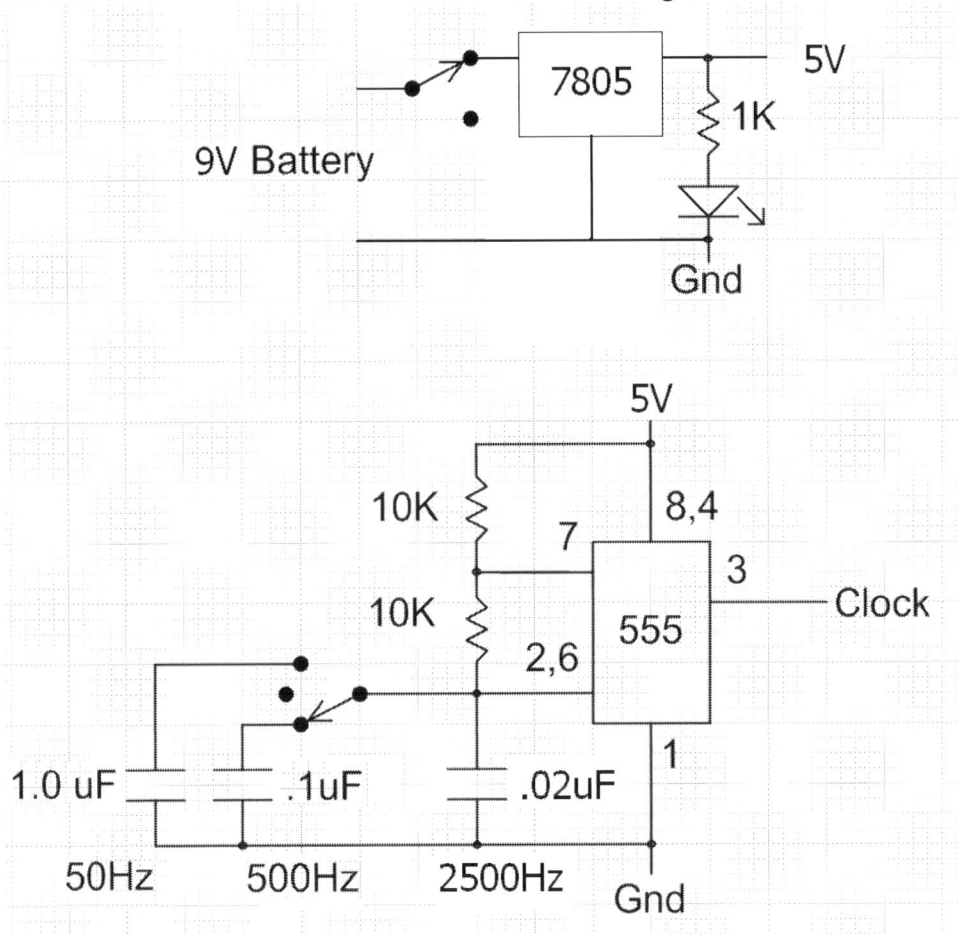

Chapter 7

LinuxCNC Program

When I first tried using LinuxCNC I could not get anything to work. When the settings were checked, they were all wrong but it would not allow you to change them. The secret is to run the "LinuxCNC Stepconf wizard" first. This wizard will then configure LinuxCNC and create a desktop icon that will then start the program up with the needed configuration.

For most of the screens in the stepper setup wizard you can just click "Forward". The second screen asks to make a new configuration and a desktop icon. Just click "Forward" unless you are modifying a configuration.

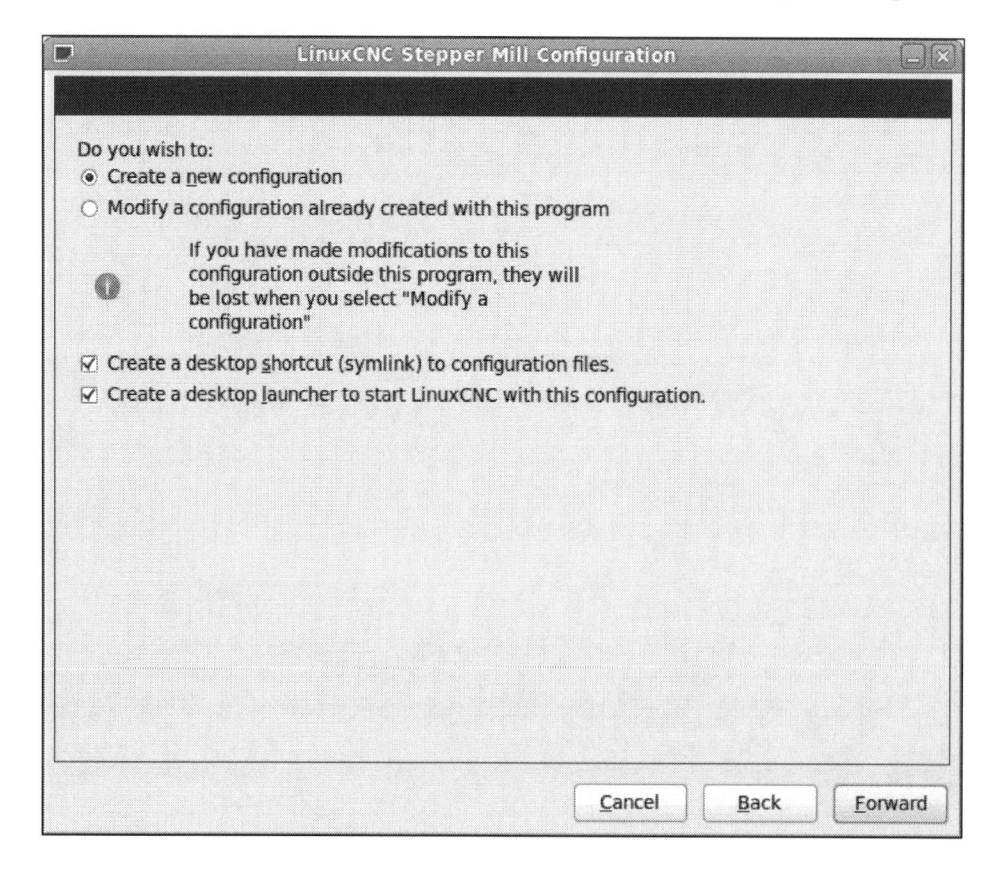

In the third screen make sure that you are using machine units of "inches" and that you are using the printer port at 0x378. At this screen, you can also rename your desktop icon or machine name and it is a good idea to run the jitter test. My PC required a higher jitter setting.

Basic machine information

Machine Name:	my-mill
Configuration directory:	~/linuxcnc/configs/my-mill
Axis configuration:	XYZ
Machine units:	Inch

Driver characteristics: (Multiply by 1000 for times specified in μs or microseconds) Additional signal conditioning or isolation such as optocouplers and RC filters can impose timing constraints of their own, in addition to those of the driver.

Driver type: Other

▽ Driver Timing Settings

Step Time:	10000	ns
Step Space:	10000	ns
Direction Hold:	200000	ns
Direction Setup:	200000	ns

▽ Parallel Port Settings

First Parport Base Address:	0x378	Out
☐ Second Parport Address:	Enter Address	Out
☐ Third Parport Address:	Enter Address	Out

Base Period Maximum Jitter: 19000 ns   Min Base Period: 29000 ns

☑ Onscreen prompt for tool change   Test Base Period Jitter   Max step rate: 17241 Hz

[Cancel] [Back] [Forward]

Just forward over the next screen. Then you should get the "Parallel port setup", one of the most critical screens of all. Make sure your pin assignments are correct and do not click on the "invert" box for pin 10 through pin 12 even though they are active when low. If you do click on invert they will work backwards! Even if I clicked on the invert for the "ESTOP Out" it did not allow me to run the axis tests in the next screens as the E-Stop seemed to still be on. I discovered that the optocouplers are over sensitive. The solution is to add a 220 or 470 ohm resistor in series with the enable signal.

Parallel Port Setup

Outputs (PC to Mill):	Invert	Inputs (Mill to PC):	Invert
Pin 1: ESTOP Out	☐	Pin 10: Home X	☐
Pin 2: X Step	☐	Pin 11: Home Y	☐
Pin 3: X Direction	☐	Pin 12: Unused	☐
Pin 4: Y Step	☐	Pin 13: Unused	☐
Pin 5: Y Direction	☐	Pin 15: Unused	☐
Pin 6: Z Step	☐		
Pin 7: Z Direction	☐		
Pin 8: A Step	☐		
Pin 9: A Direction	☐		
Pin 14: Spindle CW	☐	Output pinout presets:	
Pin 16: Spindle PWM	☐	Sherline Outputs	
Pin 17: Amplifier Enable	☐	Xylotex Outputs	

[Cancel]   [Back]   [Forward]

The next screen pictures are the setup screens for each of the three axes. I initially set the leadscrew pitch to "10" as in 10 threads per inch. I measured it with a tape measure; there are 10 grooves or tracks per inch. Then when testing the CNC, it moved four inches for a setting of two inches. Looking closely at the leadscrew I discovered that it has two starts or tracks. Because of that the correct pitch setting was actually "five".

Select "Test this axis" and a smaller dialog box should come up. In the smaller box you can adjust and then test your settings. At first I had to set the velocity all the way down to .1 or .2 to get the CNC axis to work smoothly. Then eventually I learned that you have to set the step and direction times higher under basic machine information. Now the velocity works at .8 to even 1.0.

Once you have it working set the test area to something reasonable like two or three inches for the X and Y axis. Then select "run" and see if it moves smoothly back and forth and if it always returns to the same spot. If so you now have it working. Note that my machine moves four inches for a setting of two inches but then when I run Linux CNC the distance is actually correct.

You will need to do the same thing for each axis except for when you are testing the Z axis set the test range to only one or two inches.

This is the X axis setup screen. I set the table travel to 11 inches. You can leave the home switch at "0" and it will still work.

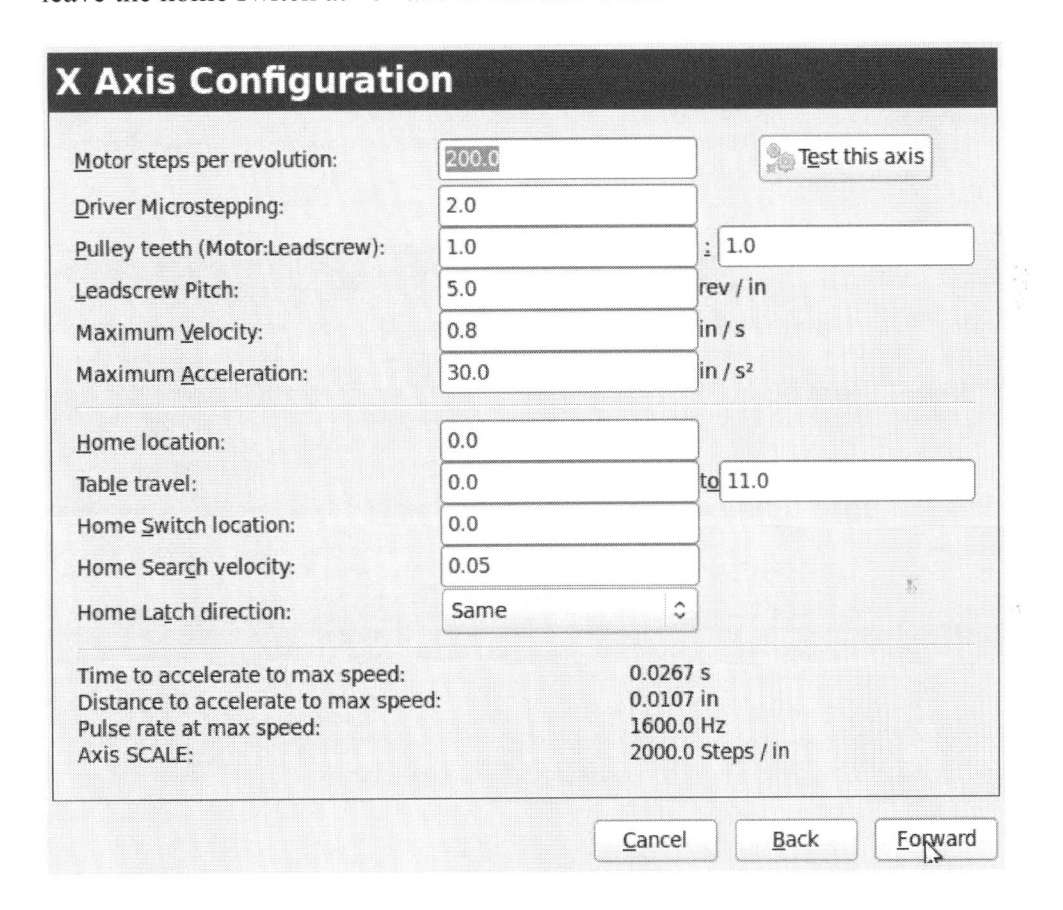

X Axis Configuration

Motor steps per revolution:	200.0	
Driver Microstepping:	2.0	
Pulley teeth (Motor:Leadscrew):	1.0 : 1.0	
Leadscrew Pitch:	5.0	rev / in
Maximum Velocity:	0.8	in / s
Maximum Acceleration:	30.0	in / s²
Home location:	0.0	
Table travel:	0.0 to 11.0	
Home Switch location:	0.0	
Home Search velocity:	0.05	
Home Latch direction:	Same	

Time to accelerate to max speed: 0.0267 s
Distance to accelerate to max speed: 0.0107 in
Pulse rate at max speed: 1600.0 Hz
Axis SCALE: 2000.0 Steps / in

[Cancel] [Back] [Forward]

Here is a picture of the pop up "test this axis" box.

73

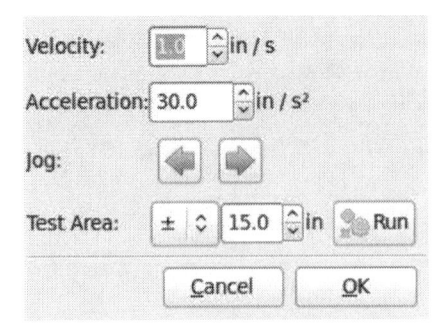

Velocity:	1.0	in / s
Acceleration:	30.0	in / s²
Jog:	← →	
Test Area:	± 15.0 in	Run
	Cancel OK	

This is the setup screen for the Y axis. The table travel could be set to eight or nine inches. When I have the wooden crash board attached, it limits my Y axis travel to only about 6 inches. Home can be 0 and velocity works as high as 1, but you should use something less to be on the safe side.

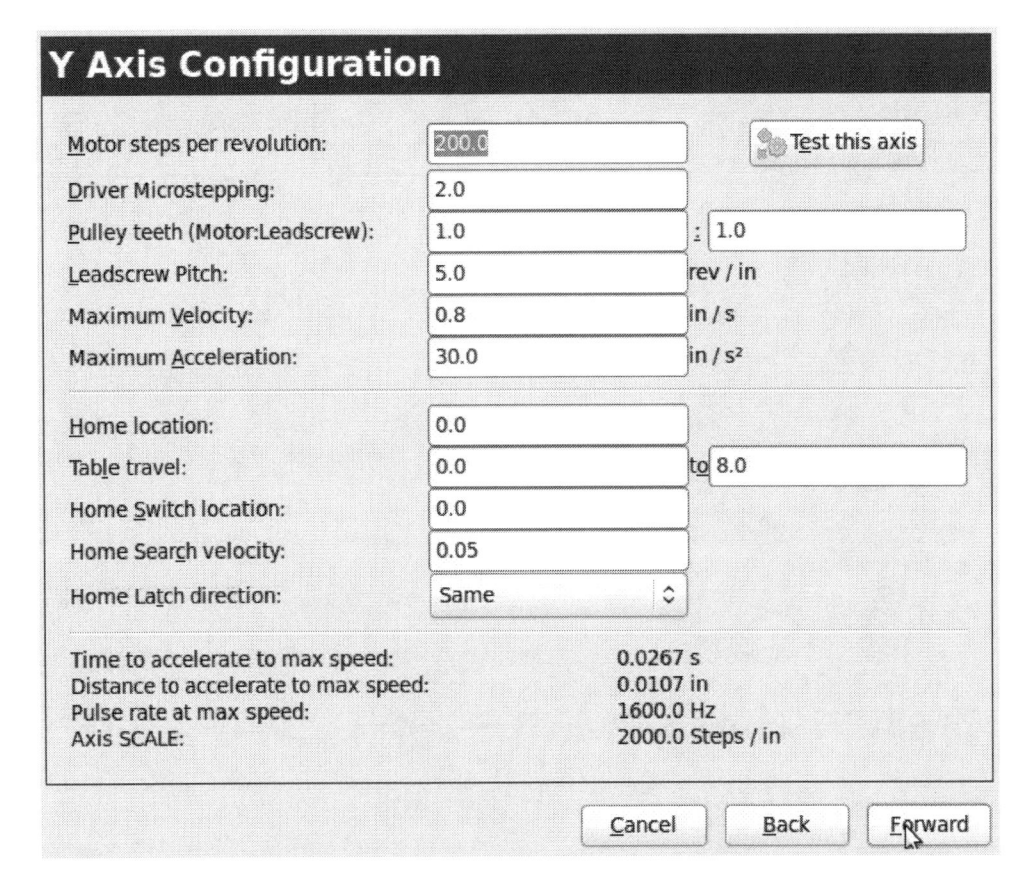

Y Axis Configuration

Motor steps per revolution:	200.0	Test this axis
Driver Microstepping:	2.0	
Pulley teeth (Motor:Leadscrew):	1.0	: 1.0
Leadscrew Pitch:	5.0	rev / in
Maximum Velocity:	0.8	in / s
Maximum Acceleration:	30.0	in / s²
Home location:	0.0	
Table travel:	0.0	to 8.0
Home Switch location:	0.0	
Home Search velocity:	0.05	
Home Latch direction:	Same	

Time to accelerate to max speed: 0.0267 s
Distance to accelerate to max speed: 0.0107 in
Pulse rate at max speed: 1600.0 Hz
Axis SCALE: 2000.0 Steps / in

Cancel   Back   Forward

Finally, there is the setup screen for the Z axis. Note that the default table travel is "-4" at the bottom going up to "0" at the top. My machine can do -7 inches. However, you usually manually "home" the Z axis to just above the surface of what you are milling.

Z Axis Configuration

Motor steps per revolution:	200.0	Test this axis
Driver Microstepping:	2.0	
Pulley teeth (Motor:Leadscrew):	1.0	: 1.0
Leadscrew Pitch:	5.0	rev / in
Maximum Velocity:	0.8	in / s
Maximum Acceleration:	30.0	in / s²
Home location:	0.0	
Table travel:	-4.0	to 0.0
Home Switch location:	0.0	
Home Search velocity:	0.05	
Home Latch direction:	Same	

Time to accelerate to max speed: 0.0267 s
Distance to accelerate to max speed: 0.0107 in
Pulse rate at max speed: 1600.0 Hz
Axis SCALE: 2000.0 Steps / in

Cancel Back Forward

Now you should be ready to run Linux CNC! When you load the program, it comes up with a test that should print out "LinuxCNC" on a piece of paper. There are other sample programs that can draw a snowflake or a spiraling circle among many other things.

Load a marker or pen into the tool area, I rubber-banded the marker to the Dremel for this test. Using the shift and arrow keys set the CNC to the front left corner with the Z axis down so that the pen just touches the paper. Turn off the "E" stop (The red "X"). Then tell it that "home" is set for all three axes. To do that, select the axis then click on "home". Touch off or offset the Z axis by about one inch and then select "Run".

Here is a picture of the Linux CNC opening screen.

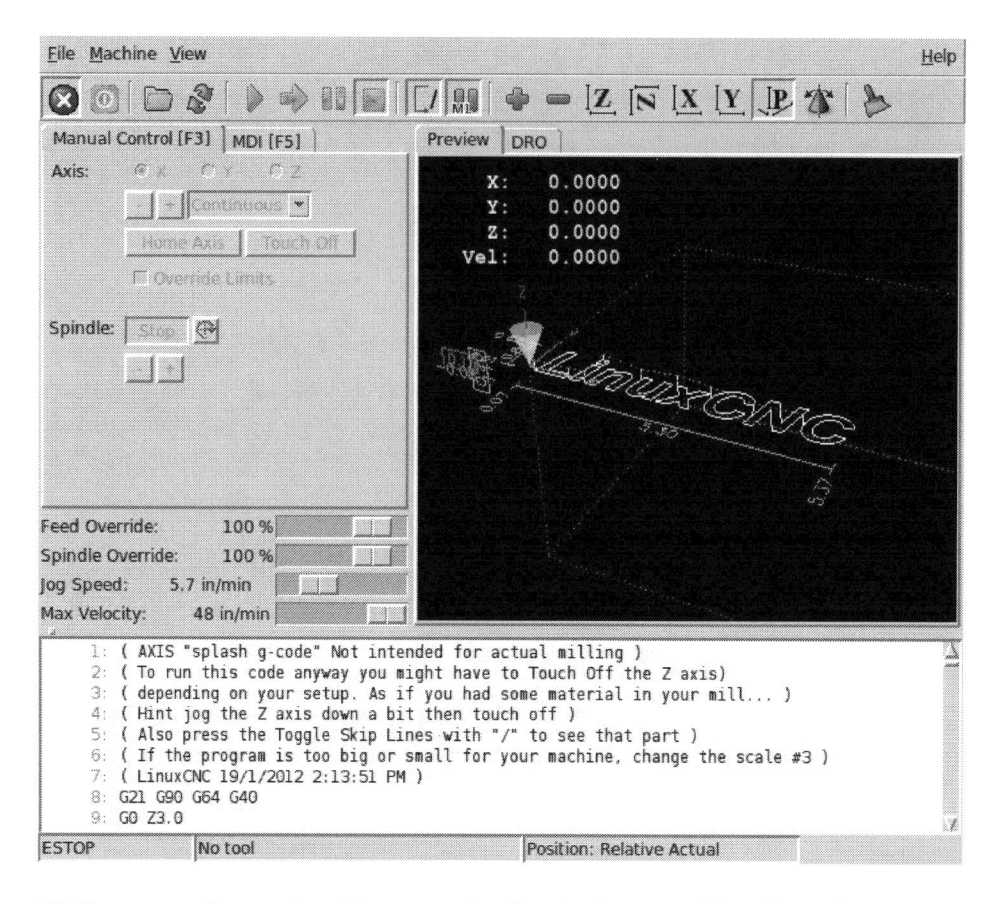

If all goes well you should get results that look something like what the following picture shows. Your marker will move around a lot if you use rubber bands, the pen needs to be clamped in place.

When I first ran this test, the Z axis was not coming down far enough each time so I had to manually help it out. So the beginning of each letter was faint or missing. I eventually purchased another stepper controller for the Z axis and now it runs up and down very smoothly.

Here is a picture of a test run with the Dremel cutting into some plastic covered chipboard. I think my feed rate was too high so it forced it to move faster than it was able to cut resulting in some rough edges.

The Linux CNC menu or tool bar could use some additional explanation as to what the many icons do. Here is a simplified chart taken from the Linux CNC manual.

```
The toolbar buttons from left to right in the Axis
display. Keyboard shortcuts are shown [in brackets]
```

- ⊗ Toggle Emergency Stop [F1] (E-Stop)
- ⓘ Toggle Machine Power [F2]
- 📂 Open G Code file [O]
- 🔄 Reload current file [Ctrl-R]
- ▶ Begin executing the current file [R]
- ➡ Execute next line [T]
- ⏸ Pause Execution [P] Resume Execution [S]
- ⏹ Stop Program Execution [ESC]
- Toggle Skip lines with "/" [Alt-M-/]
- Toggle Optional Pause [Alt-M-1]
- ✚ Zoom In   ➖ Zoom Out
- [Z] Top view   [N] Rotated Top view
- [X] Side view   [Y] Front view
- [P] Perspective view
- 🧭 Toggle between Drag and Rotate Mode [D]
- ✎ Clear live backplot [Ctrl-K]

Linux CNC also has some "hidden" keyboard shortcuts. These do things like give you the ability to move each axis around rapidly. I was using my hand or the pendant to make some position adjustments until I found out about these keyboard commands.

```
Escape          Stop motion
F1              Toggle estop/estop reset state
F2              Toggle machine off/machine on state
F3              Manual mode
F4              Auto mode
X, `            Select X axis
Y, 1            Select Y axis
Z, 2            Select Z axis
Left, Right     Jog X axis (Shift accelerates)
Up, Down        Jog Y axis (Shift accelerates)
Page Up, Down   Jog Z axis (Shift accelerates)
Home            Home selected axis
End             Touch off selected axis
<, >            Decrease or increase axis speed
C               Select continuous jogging
I               Select jog increments
1-9,0           Set feed override to 10% to 100%
O               Open a program
R               Run an opened program
P               Pause an executing program
S               Resume a paused program
A               Step one line in a paused program
```

An interesting feature hidden in LinuxCNC is a picture to CNC code converter. It does take some time to do but the results are very interesting. What it does is to break down each row of pixels in the picture into a horizontal scan of the CNC machine. The depth of the Z axis while scanning across the picture is related to the darkness of the image. By using this feature, you could possibly take a circuit board layout made in PC paint, or that you found on the web someplace and convert it into a CNC program to produce that circuit board.

You could also use this feature to make a picture of a company logo into a Plexiglas sign by milling the picture into the Plexiglas and then lighting it up with some LED's mounted in a wood frame for the Plexiglas.

Chapter 8

CNC Codes

I have created three one page lists of the CNC commands, G codes, and M codes. Well, except for the G codes, they took two pages.

A – Position of A axis
B – Position of B Axis
C – Position of C Axis
D – Diameter or Depth of cut
E – Precision feed rate
F – Feed rate
G – Motion commands 1-99
H – Tool length offset
I – Arc Center for X
J – Arc Center for Y
K – Arc Center for Z
L – Fixed cycle loop
M - Miscellaneous function
N – Line number
O – Program name
P – Parameter Address
Q – Peck increment
R – Size of arc radius or retract height
S – Speed
T – Tool selection
U – Incremental axis corresponding to X
V – Incremental axis corresponding to Y
W – Incremental axis corresponding to Z
X – Absolute or incremental position of X
Y – Absolute or incremental position of Y
Z – Absolute or incremental position of Z

G Codes expanded

G00 - Rapid positioning
G01 – Linear Interpolation
G02 - Circular interpolation, clockwise
G03 - Circular interpolation, counterclockwise
G04 – Dwell
G05 - High-precision contour control
G06 - Non Uniform Rational B Spline
G07 – Imaginary Axis
G09 - Exact stop check
G10 – Programmable data input
G11 - Data write cancel
G12 - Full-circle interpolation, clockwise
G13 - Full-circle interpolation, counterclockwise
G17 - XY plane selection
G18 - ZX plane selection
G19 - YZ plane selection
G20 - Programming in inches
G21 - Programming in millimeters
G28 – Return to home
G30 – Return to secondary home
G31 – Skip
G32 – Single point threading
G33 – Constant pitch threading
G34 – Variable pitch threading
G40 - Tool radius compensation off
G41 - Tool radius compensation left
G42 - Tool radius compensation right
G43 - Tool radius compensation negative
G44 - Tool radius compensation positive
G45 - Axis offset single increase
G46 - Axis offset single decrease
G47 - Axis offset double increase
G48 - Axis offset double decrease
G49 - Tool length offset compensation cancel
G50 - Define the maximum spindle speed
G50 - Scaling function cancel
G50 - Position register (vector from part zero to tool tip)
G52 - Local coordinate system
G53 - Machine coordinate system
G54 - G59 Work coordinate systems

G61 - Exact stop check, modal
G62 - Automatic corner override
G64 - Default cutting mode - cancel exact stop check mode
G70 - Fixed cycle, multiple repetitive cycle, for finishing
G71 - Fixed cycle, multiple repetitive cycle, for roughing Z-axis
G72 - Fixed cycle, multiple repetitive cycle, for roughing X-axis
G73 - Fixed cycle, multiple repetitive cycle, for roughing pattern
G73 - Peck drilling cycle for milling – NO full retraction
G74 - Peck drilling cycle for turning
G74 - Tapping cycle for milling left hand thread
G75 - Peck grooving cycle for turning
G76 - Fine boring cycle for milling
G76 - Threading cycle for turning repetitive cycle
G80 - Cancel canned cycle
G81 - Simple drilling cycle
G82 - Drilling cycle with dwell
G83 - Peck drilling cycle full retraction
G84 - Tapping cycle right hand thread
G85 - Boring cycle feed in/feed out
G86 - Boring cycle feed in/spindle stop/rapid out
G87 - Boring cycle back boring
G88 - Boring cycle feed in/spindle stop/manual operation
G89 - Boring cycle feed in/dwell/feed out
G90 - Absolute programming
G90 - Fixed cycle, simple cycle, for roughing Z-axis
G91 - Incremental programming
G92 - Position register vector from part zero to tool tip
G92 - Threading cycle, simple cycle
G94 - Feed rate per minute
G94 - Fixed cycle, simple cycle, for roughing X-axis
G95 - Feed rate per revolution
G96 - Constant surface speed
G97 - Constant spindle speed
G98 - Return to initial Z level in canned cycle
G98 – Feed rate per minute group type A
G99 - Return to R level in a canned cycle
G99 – Feed rate per revolution group type A

M codes expanded

M00 - Compulsory stop
M01 - Optional stop
M02 - End of program
M03 - Spindle on clockwise rotation
M04 - Spindle on counterclockwise rotation
M05 - Spindle stop
M06 - Automatic tool change
M07 - Coolant on mist
M08 - Coolant on flood
M09 - Coolant off
M10 - Pallet clamp on
M11 - Pallet clamp off
M13 - Spindle on clockwise rotation and coolant on flood
M19 - Spindle orientation
M21 - Mirror X axis M21 - Tailstock forward
M22 - Mirror Y axis M22 - Tailstock backward
M23 - Mirror off
M23 - Thread gradual pullout on
M24 - Thread gradual pullout off
M30 - End of program return to program top
M41 - Gear 1 select
M42 - Gear 2 select
M43 - Gear 3 select
M44 - Gear 4 select
M48 - Feed rate override is allowed
M49 - Feed rate override is not allowed
M52 - Unload last tool from spindle
M60 - Automatic pallet change
M98 - Subprogram call
M99 - Subprogram end

One thing that makes writing G Code easier to work with is a program that allows you to see what your code does as you type it. I found a nice G code emulator on line that is written in HTML called "SimpleGcoder". It is found at simplegcoder.com. Since I program in a number of languages, I know that doing this in HTML is quite an accomplishment. You can even right click, reveal code, copy the code and paste it into notepad, then save it as SimpleG Coder.html. Then you can run the program off line!

One of the drawbacks of using simple G coder is that all circles have to be programmed using the radius command. Here is a screen capture of it.

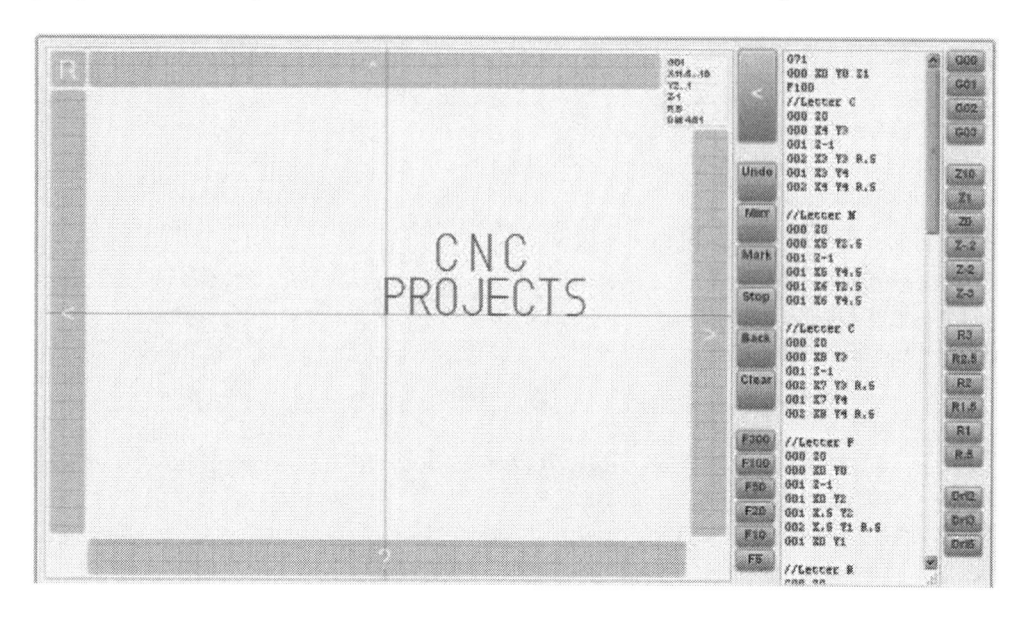

Here is a sample CNC program that I wrote to print or engrave "CNC PROJECTS" into wood. The Z axis is set to .9 inches above the wood to be engraved to cut into it to a 0.1 inch depth.

G20   // Programming in inches
G64   // Default cutter
G17   // XY Plane
G40   // Tool Comp off
G00 X0 Y0 Z0 //Home
(D.2)   //Diameter of cut
F100   // Feed rate

//Letter C
G00 Z0
G00 X4 Y3
G01 Z-1
G02 X3 Y3 R.5
G01 X3 Y4
G02 X4 Y4 R.5

//Letter N
G00 Z0
G00 X4.5 Y2.5

G01 Z-1
G01 X4.5 Y4.5
G01 X5.5 Y2.5
G01 X5.5 Y4.5

//Letter C
G00 Z0
G00 X7 Y3
G01 Z-1
G02 X6 Y3 R.5
G01 X6 Y4
G02 X7 Y4 R.5

//Letter P
G00 Z0
G00 X0 Y0
G01 Z-1
G01 X0 Y2
G01 X.5 Y2
G02 X.5 Y1 R.5
G01 X0 Y1

//Letter R
G00 Z0
G00 X1.25 Y0
G01 Z-1
G01 X1.25 Y2
G01 X1.75 Y2
G02 X1.75 Y1 R.5
G01 X1.25 Y1
G01 X1.75 Y1
G01 X2.25 Y0

//Letter O
G00 Z0
G00 X3.5 Y.5
G01 Z-1
G02 X2.5 Y.5 R.5
G01 X2.5 Y1.5
G02 X3.5 Y1.5 R.5
G01 X3.5 Y.5

//Letter J
G00 Z0
G00 X3.75 Y.5
G01 Z-1
G03 X4.75 Y.5 R.5
G01 X4.75 Y.5
G01 X4.75 Y2

//Letter E
G00 Z0
G00 X5 Y0
G01 Z-1
G01 X5 Y2
G01 X6 Y2
G01 X5 Y2
G01 X5 Y1
G01 X6 Y1
G01 X5 Y1
G01 X5 Y0
G01 X6 Y0

//Letter C
G00 Z0
G00 X7.25 Y.5
G01 Z-1
G02 X6.25 Y.5 R.5
G01 X6.25 Y1.5
G02 X7.25 Y1.5 R.5

//Letter T
G00 Z0
G00 X8 Y0
G01 Z-1
G01 X8 Y2
G01 X7.5 Y2
G01 X8.5 Y2

//Letter S
G00 Z0
G00 X8.75 Y0
G01 Z-1
G01 X9.25

```
G03 X9.25 Y1 R.5
G02 X9.25 Y2 R.5
G01 X9.75

G00 Z0 // Raise Bit
M2  // end of program
```

Chapter 9

Router Upgrade

When using the Dremel I had to reduce the feed rate to about 10% or the bit would start smoking. I needed something much more powerful. I had purchased a router a few years ago for $20 at a yard sale. The handles unscrewed from the router leaving a 3.5 inch diameter area to use to mount the router.

However the hardware stores did not have a pipe clamp that big. I purchased a two inch pipe clamp and used some vice grips and pliers to enlarge it to 3.5 inches. Behind the router I installed a rubber "foot" about 1 inch in diameter and .5 inches tall. The pipe strap was short about 1.5 inches so I added spacers on each side. I used 8-32 by 2 inch screws and nuts to go from the strap through the spacers and through the Z axis platform to hold it in place.

Here is a picture of the router mounting hardware with one screw removed.

With the router installed I was able to cut much deeper and do that with a feed rate set to 100%. I thought you could adjust the feed rate while it was running and doing that crashed the Z axis. The crash put a hole at the bottom of the "N" in this picture.

The next picture is of some of some of the router bits. They only have two teeth typically so they can remove more wood per cycle, and as a result these bits require a faster rotational speed.

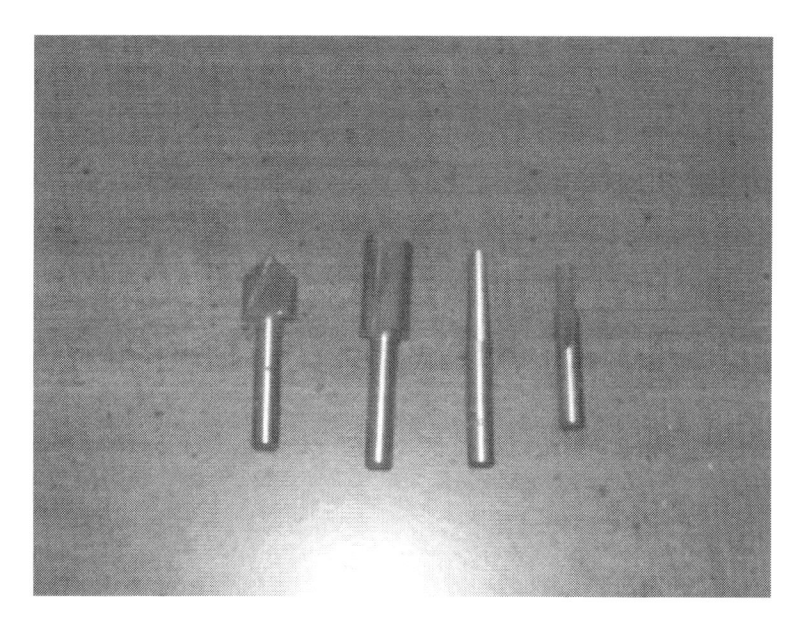

Here is the drawing for the Z axis platform modified to fit the router instead of the Dremel.

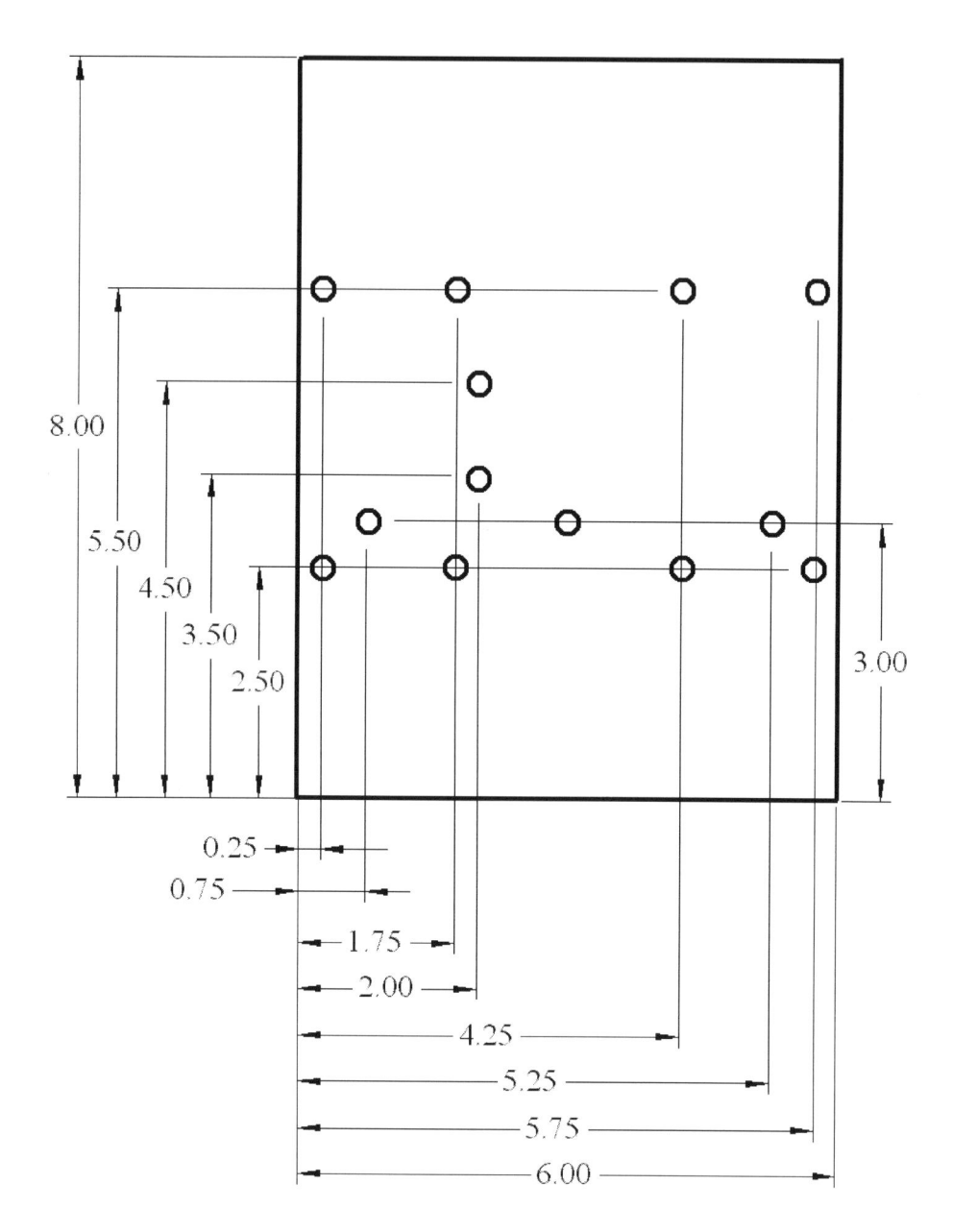

Chapter 10

Metal Frame Upgrade

Someone who saw my home-made CNC machine, when it was made out of Plexiglas, gave me some aluminum pieces to make it into a better machine. The aluminum pieces consisted of six pieces of three inch by one inch by 18-inch-long "C" Channel and some flat pieces of aluminum as well. All the metal pieces were .125 inch thick. With those pieces, I designed a new base frame as well as new gantry sides. This new design does not have a solid base under it. It is basically a square frame with the two gantry sides bolted onto it.

The corners of the frame could be either welded or bolted together. Since I did not have a welder or know how to weld, I elected to bolt it together instead. I decided to use Hexagonal 1/4-28 bolts and nuts because the bolt heads also make nice legs for the machine to sit on. Welding is a better method because my machine keeps getting out of square and needs adjustment before it is used.

You could use three inch by one inch "C" channel for the sides as well as the front and back but the top edge might get in the way of the Y axis platform or carriage assembly. To fix that problem you would make the carriage assembly just a little smaller so it would clear, or you could cut the top edge of the three by one inch channel down to where it is only about 1/2 of an inch wide.

Up next is a mechanical drawing of the metal parts layout as it is seen from the front of the machine.

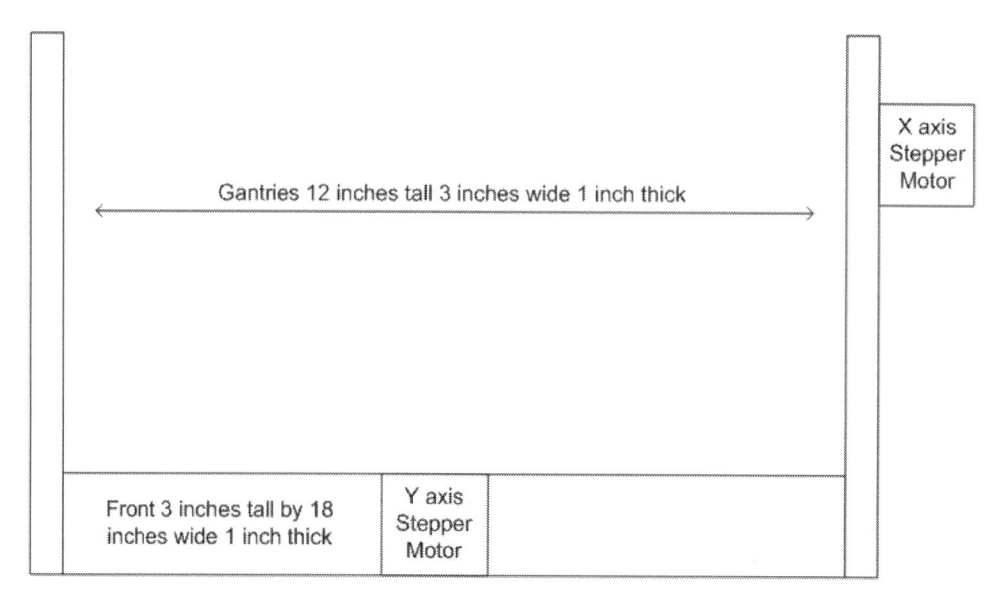

```
|                                                          | X axis  |
|                                                          | Stepper |
|                                                          | Motor   |
  Gantries 12 inches tall 3 inches wide 1 inch thick
```

```
| Front 3 inches tall by 18 | Y axis  |
| inches wide 1 inch thick  | Stepper |
|                           | Motor   |
```

I did not want to have to drill any 1/2 inch diameter holes into the aluminum so I opted to use 1/2 inch rod holders instead. However, only metric rod holders are commonly available so I ordered eight 13 mm rod holders. There needs to be two for both ends of the two X axis guide rods and two for both ends of the Y axis guide rods. These rod holders will also hold the guide rods in place much better than the holes in the Plexiglas did. Putting a layer of electrical tape on the rods before inserting them in the holders will make up for the minor size difference.

The hole to hole spacing for mounting the shaft holders is about 1.5 inches or about 3/4 an inch from each side of the center mark for the rod holes. The exact spacing is 36mm or 1.4 inches hence .7 inches on either side of the center mark. I used .75 inches for my machine and ended up filing out some holes so the screws would fit.

Also the gantry sides are now three inches wide instead of being six inches wide. They will no longer reach all the way to the back of the machine. An alternate rear support method might be needed. One method would be to use a piece of aluminum that would fasten to the gantry and then to the back of the machine. Another possibility would be to do nothing at all because even without being fastened to the back of the machine, the gantry is still a lot more stable than when it was made out of Plexiglas.

Next there is a mechanical drawing of the new parts as seen looking down from above at the machine.

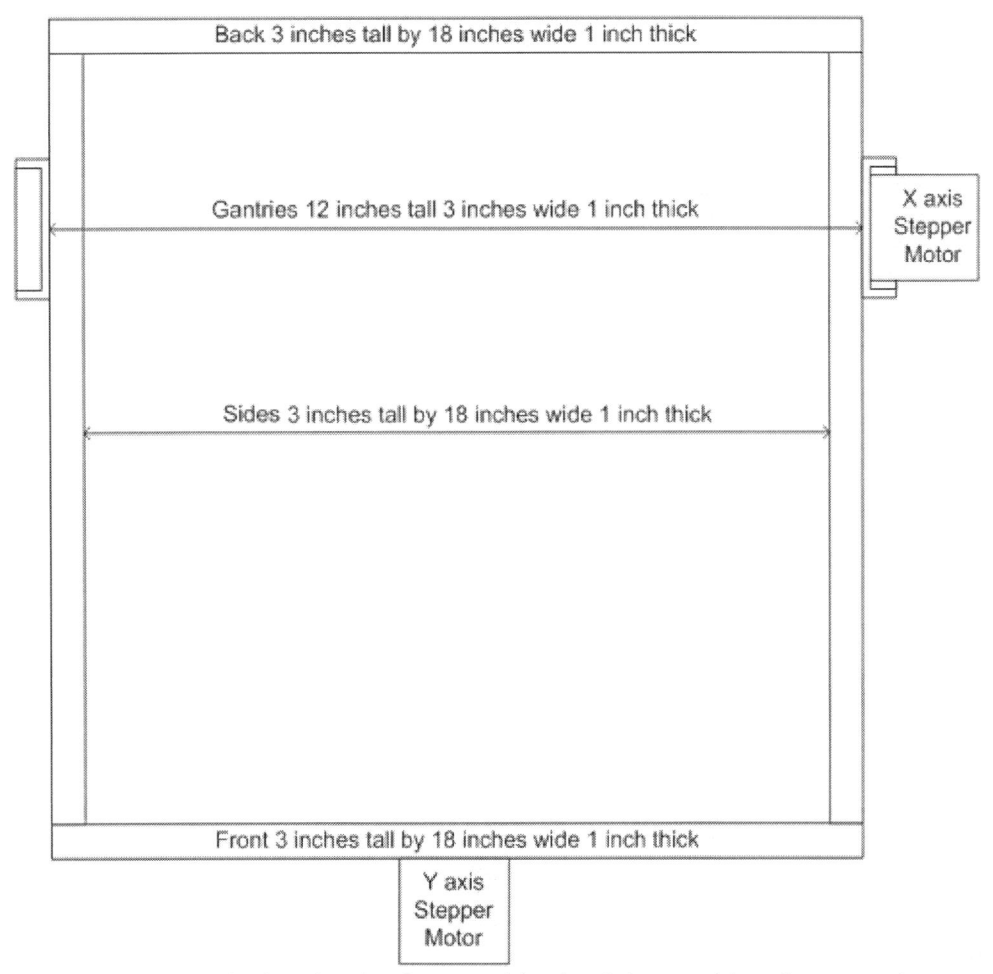

```
┌─────────────────────────────────────────────────┐
│     Back 3 inches tall by 18 inches wide 1 inch thick     │
├─────────────────────────────────────────────────┤
│                                                 │
│   ←─── Gantries 12 inches tall 3 inches wide 1 inch thick ───→ │  X axis
│                                                 │  Stepper
│                                                 │  Motor
│                                                 │
│   ←─── Sides 3 inches tall by 18 inches wide 1 inch thick ───→ │
│                                                 │
│                                                 │
├─────────────────────────────────────────────────┤
│     Front 3 inches tall by 18 inches wide 1 inch thick    │
└─────────────────────────────────────────────────┘
                     Y axis
                     Stepper
                     Motor
```

The mechanical design for the front and back of the machine does not change much. When they were made out of Plexiglas they were 18 inches by three inches and the aluminum pieces are the same size. The only change would be that the 1/2 inch holes for the support rods can be eliminated. Then at .7 inches from each side of where those holes were, a clearance hole for a 10-32 screw is drilled. The new holes are used to hold the rod holders in place.

The gantry was redesigned quite a bit, so I have included new drawings. When designing the gantry, I ran into a problem where the bearing holder does not fit between the rod holders. The solution I used was to use a capacitor mounting strap for the bearing. Even then I had to file the holes to get the screw heads to fit as close as possible to each other. Another solution would be to rotate it 90 degrees.

The next two drawings are the front and back of the CNC.

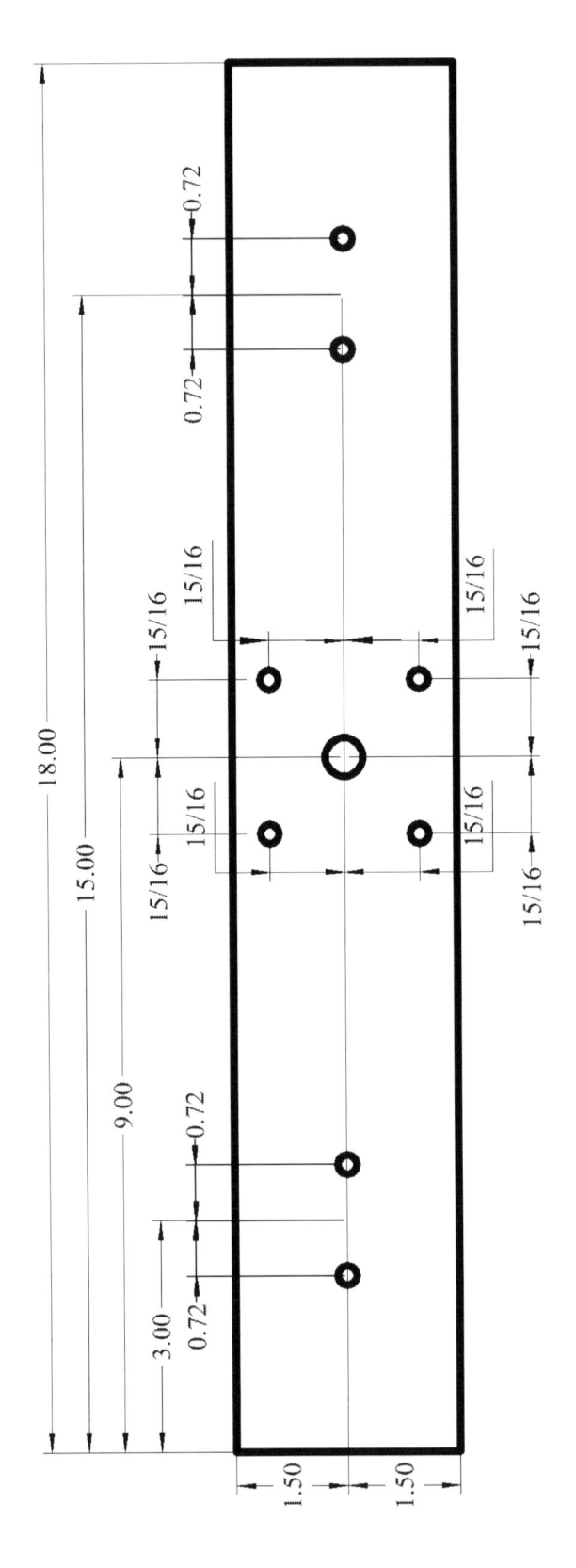

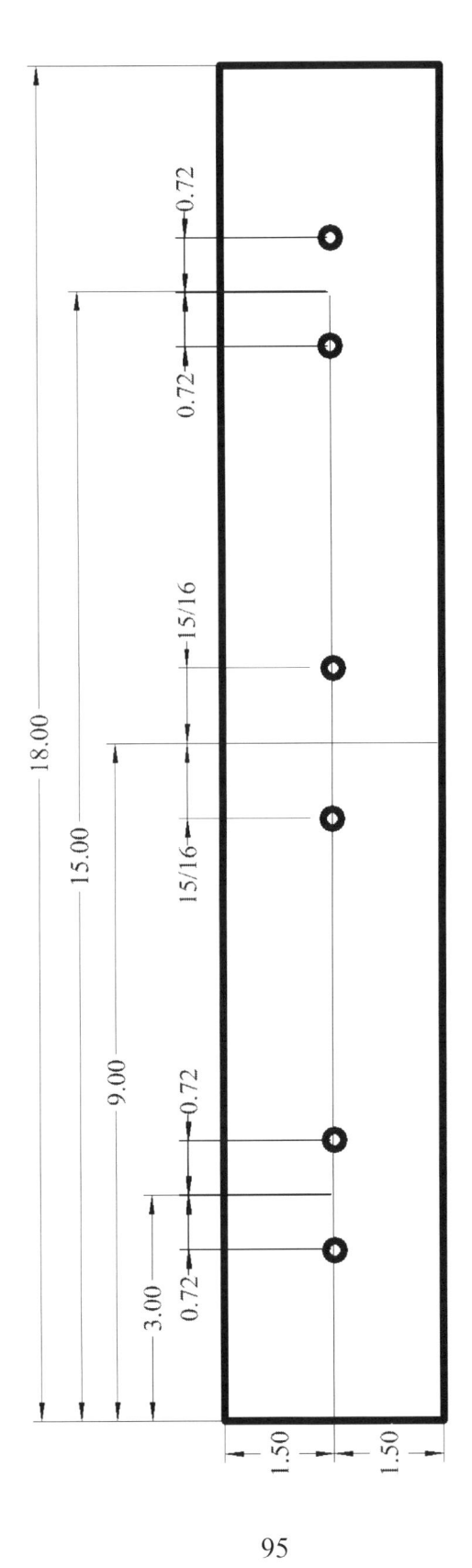

This drawing is for both sides of the CNC machine.

18.0

5.50

3.50

0.50

2.50

Here is the left side of the gantry for making it out of aluminum.

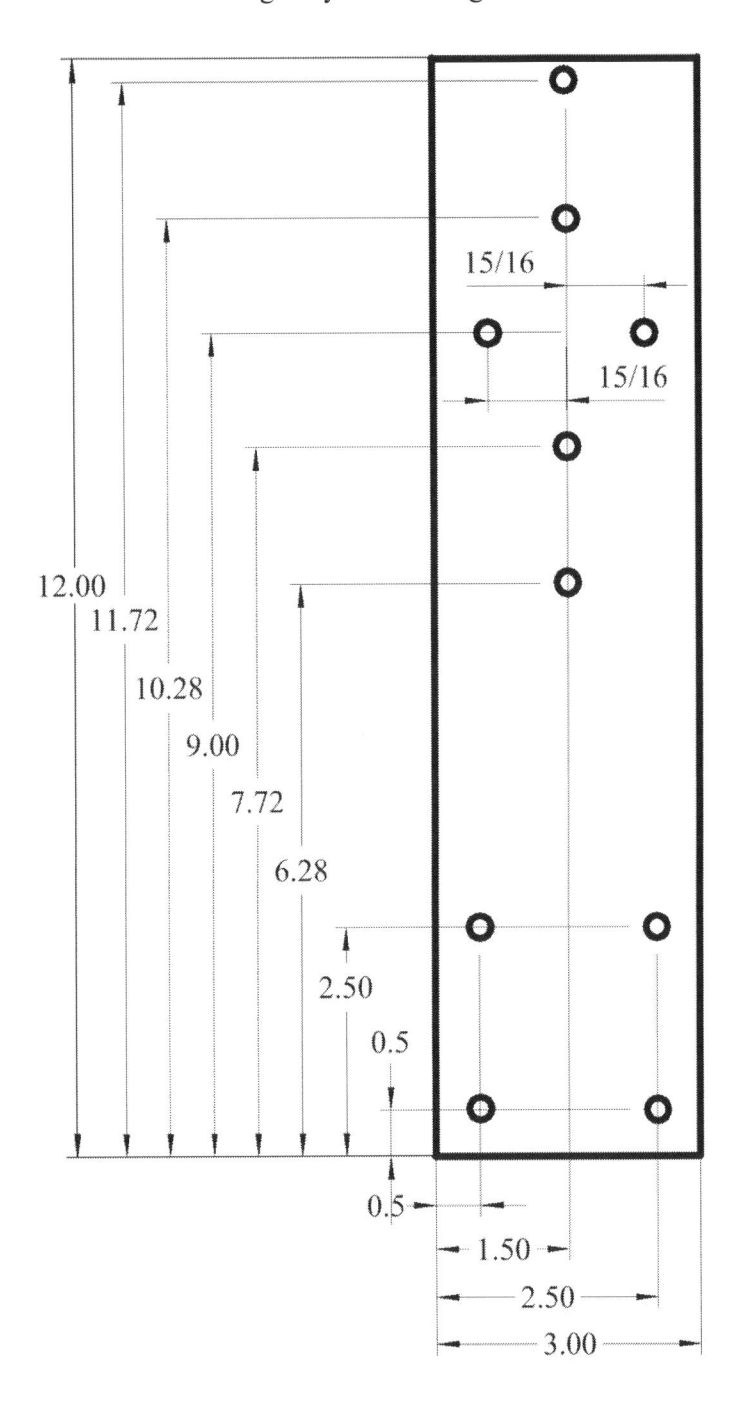

This next drawing is for the right side of the aluminum gantry.

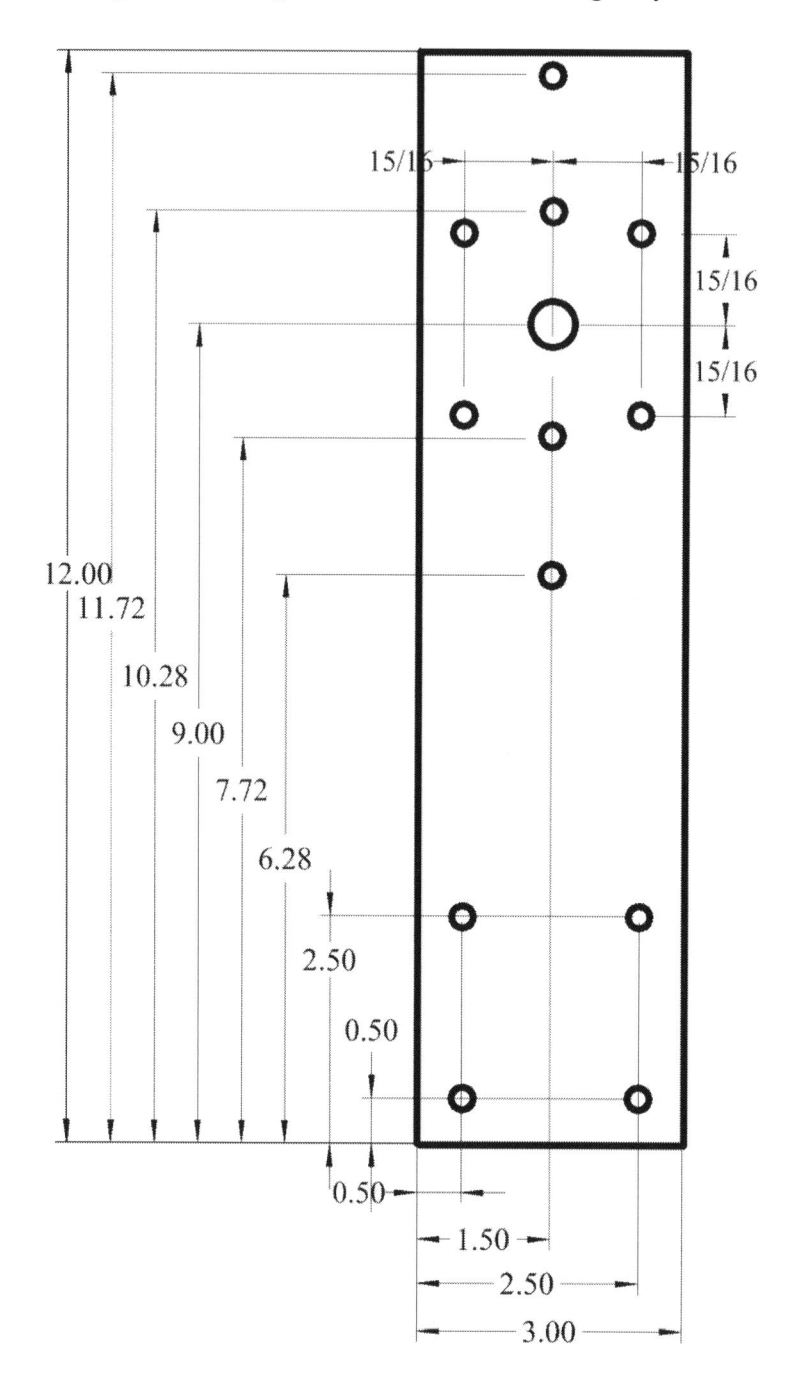

The top and bottom of the Z axis also changed. They are now 3 inches wide instead of 2 inches and have holes for rod holders. The next diagram shows how to make the Z axis top.

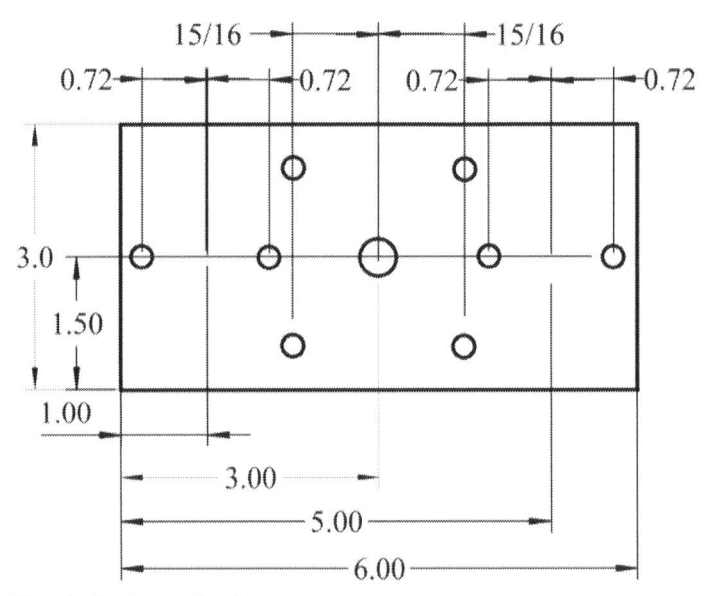

This is the Z axis bottom design.

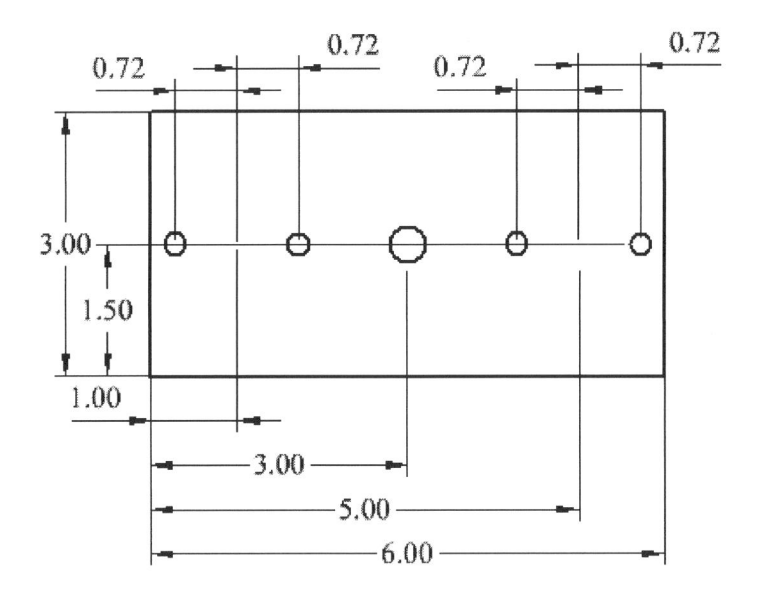

The right and left sides of the machine are 18 inches long and the gantry attaches at 3 inches from the back of the machine. They can be welded or they can have a hole at each end that lines up with a hole in the bottom corner of the front and back metal pieces. That hole should be about .5 inches by .5 inches from the lower corner of the front and back pieces.

The next picture shows the metal pieces once they are all cut and drilled. The left two pieces are actually four pieces. They are the gantry sides towards the front and the Z top and bottom towards the back. The next two pieces are the front and back. The next two pieces are the sides. The two metal plates are the X and Z axis platforms.

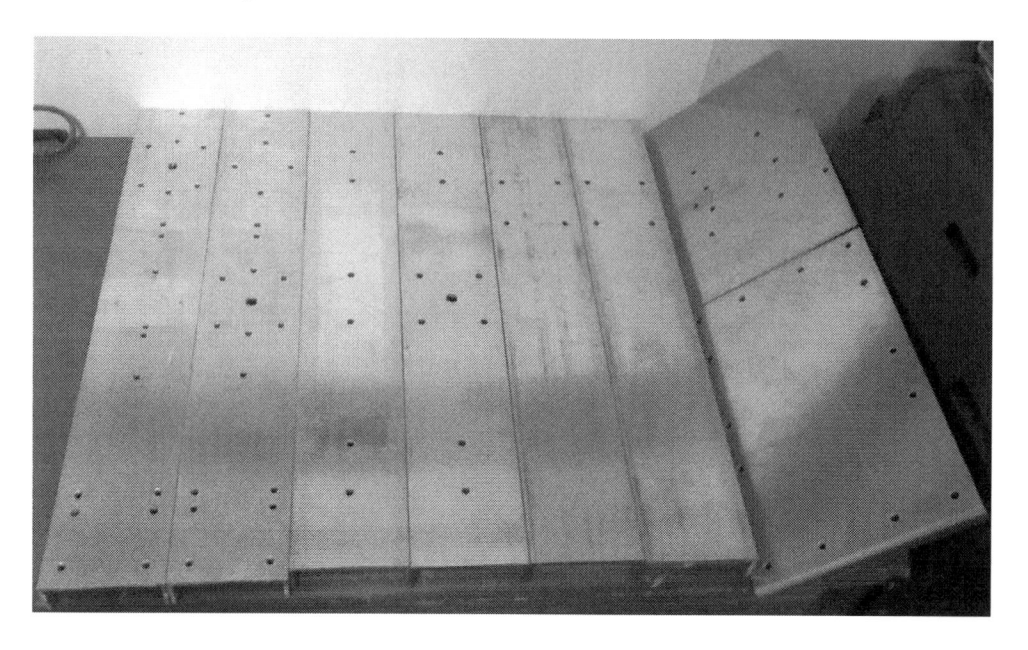

The next picture shows what the new aluminum frame looked like when it was first being put together. Be sure to make sure it is square everywhere.

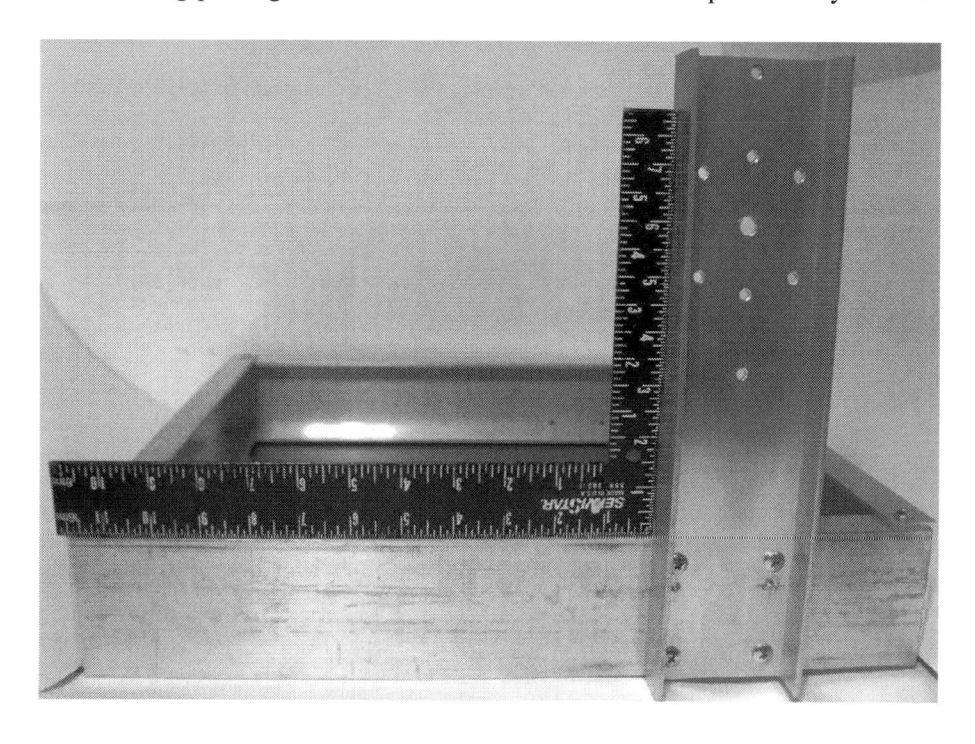

This picture shows the metal Z axis fully assembled.

Mechanical limit switches were used for the metal version of the CNC machine. Magnetic switches do not work well around metal parts. Here is a picture of the X and Y home or limit switches. They are held in place by one 4-40 screw and some glue. That way if the machine crashes into them it does not break the switches off. Switches with paddles or levers work best.

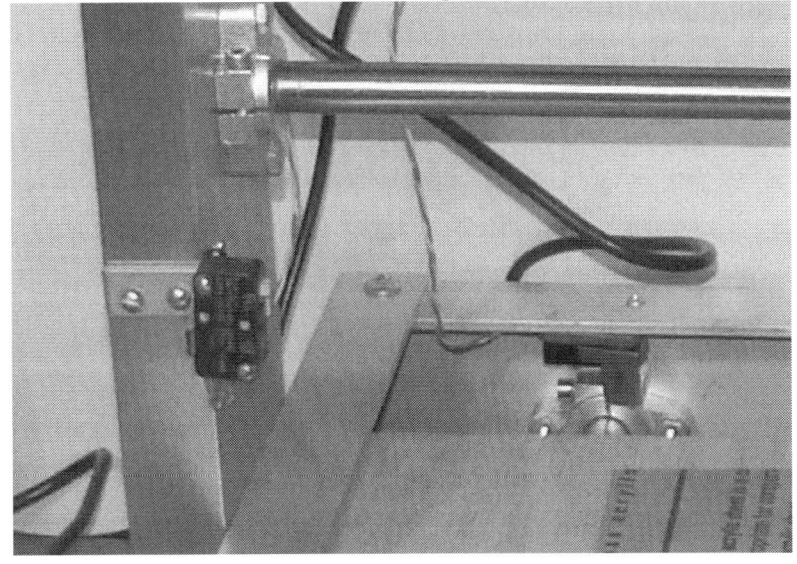

This picture is a close up that shows what the rod supports and the flexible shaft coupler looked like when they were installed on the gantry.

Some of the new parts that were used in the metal design update include these support rod holders. They replaced the rings that were used to hold the rods in place in the previous design. I tried using SHF12's but they were way too tight and had to be drilled out, that was very messy. Then I bought these SHF13's but they do not easily tighten onto the 1/2 inch shafts, so I added a layer of electrical tape to the ends of the rods and that solved the problem.

Pack of 4 SHF13 13mm Aluminum Linear Rod Rail Shaft Suppo... (300896069077)
Sale date: 08/18/14
Tracking number: LK150557474CN
Estimated Delivery: Wed. Aug. 27 - Fri. Sep. 12

$7.65
+ $1.18 shipping

An optional upgrade was to install flexible shaft couplers between the stepper motors and the lead screws. These also eliminated the need for aluminum sleeves to adapt the 5mm leadscrew shafts to the 1/4 inch stepper motor shafts.

5 x 6.35mm CNC Motor Jaw Shaft Coupler 5mm To 6.35mm Flex... (180968009041)
Sale date: 08/15/14
Tracking number: --
Estimated Delivery: Wed. Sep. 3 - Fri. Oct. 10

$2.85
Free shipping

Up next there is a picture showing the front view of the assembled Metal framed CNC machine. I am still using wood as a crash guard.

Chapter 11

Arduino GRBL Upgrade

Parallel ports on computers are going the way of the dinosaur. Unfortunately this CNC machine was designed to interface to a computer via the parallel port. A USB to printer port adapter only works with printers and will not work with a CNC machine because the timing of the commands is very critical. There are several USB to CNC interface solutions that are available. The easiest USB to CNC adapter is to make one that uses an Arduino UNO to run the CNC.

There are three ways to do the CNC conversion:

1. The first method is to make an Arduino UNO to parallel port interface adapter. This is the simplest method of conversion as you do not need to make any changes to the insides of the CNC machine.

2. The second method is to install the Arduino right into the CNC electronics. That arrangement would require that the CNC electronics are not all on one circuit board. You will need access the step and direction inputs to the stepper motor controllers.

3. The third option is to replace all of the CNC's electronics with an Arduino that is equipped with at stepper motor controller shield. This shield installs on top of the Arduino and has motor controllers that plug into it.

The Arduino UNO can run a program that is called "GRBL". The name GRBL is an abbreviated version of the Gerbil, the creature that is a little bigger than a mouse. GRBL turns the Arduino into a CNC machine controller. It uses all of the Arduino's 32K of memory and will not run on older versions that have less memory.

GRBL has changed a lot over the years and as of this writing they are on version .9. Earlier versions required a special loader to get it into the

Arduino. The current version can use the Arduino Loader. This change makes it easier to reprogram or reuse the Arduino.

To install GRBL into the Arduino do the following:
1. Unzip and Install the Arduino Interface Software.
2. Plug in the Arduino and point Windows to where you installed the driver.
3. Then unzip the GRBL zip file to a safe location.
4. Start the Arduino Interface and install the GRBL Library.
5. Select the Arduino's com port and the Arduino device version.
6. Compile and then upload GRBL to the Arduino.
7. Start the Arduino Serial monitor and select 115K baud.
8. If all went well you should see the GRBL version number.
9. Install the "Universal G Code Sender" and start it. (Requires Java)
10. Select "Machine Control" and click on X or Y +or- to see if it works.
This is what the Universal G Code Sender program looks like.

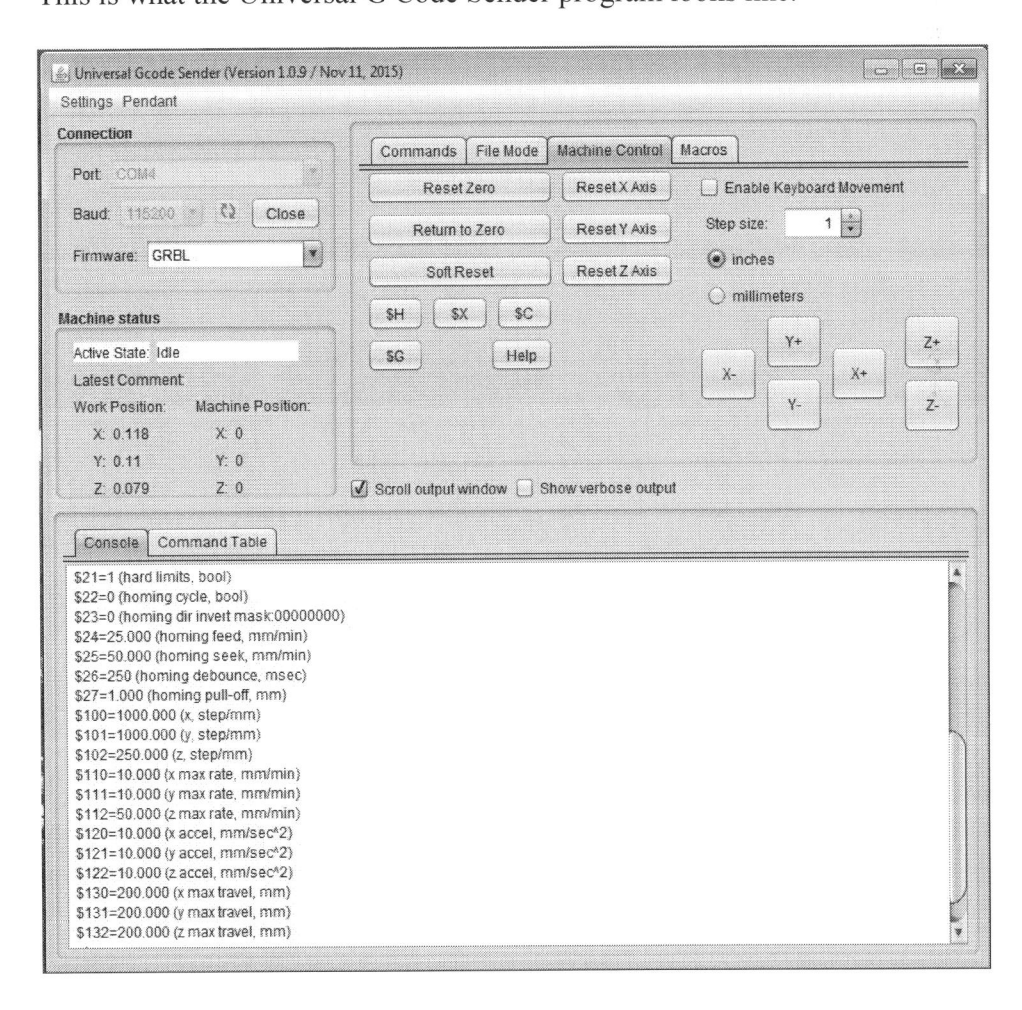

For the parallel port adapter you will need the following translation table to convert between the 25 pin printer port and the D2-D13 or data pins of the Arduino. The printer port pins will vary slightly from one CNC machine to another so be sure to check your machine setup for any pin changes.

The next chart shows one of the common pin arrangements. Pin two to nine are always output pins for the printer port (used for step and direction) and pins 10 to 13 are always input pins (Used for limit/stop switches). The Arduino pins can be either input or output pins under software control.

Printer Port Pin	Arduino Pin	Definition
1	D8	Enable
2	D2	Step X
3	D5	Direction X
4	D3	Step Y
5	D6	Direction Y
6	D4	Step Z
7	D7	Direction Z
8		Step A
9		Direction A
10	A0?	E Stop Sw
11	D9	X Limit Sw
12	D10	Y Limit Sw
13	D11	Z Limit Sw
14	D13	Spindle Dir/Cw
15	A0?	Suspend Sw
16	D12	Spindle Enable
17		Amp Enable/Hi Pwr
18-25	Gnd	Ground

This next picture shows a simple Arduino to CNC 25 pin parallel port adapter. The adapter was made with some network cables for the wires that go between the Arduino shield and the 25 pin connector. Three of the network cable pairs are "step" and "direction" for the X, Y, and Z axis. Another three of the pairs are the stop switches and ground for X, Y, and Z axis.

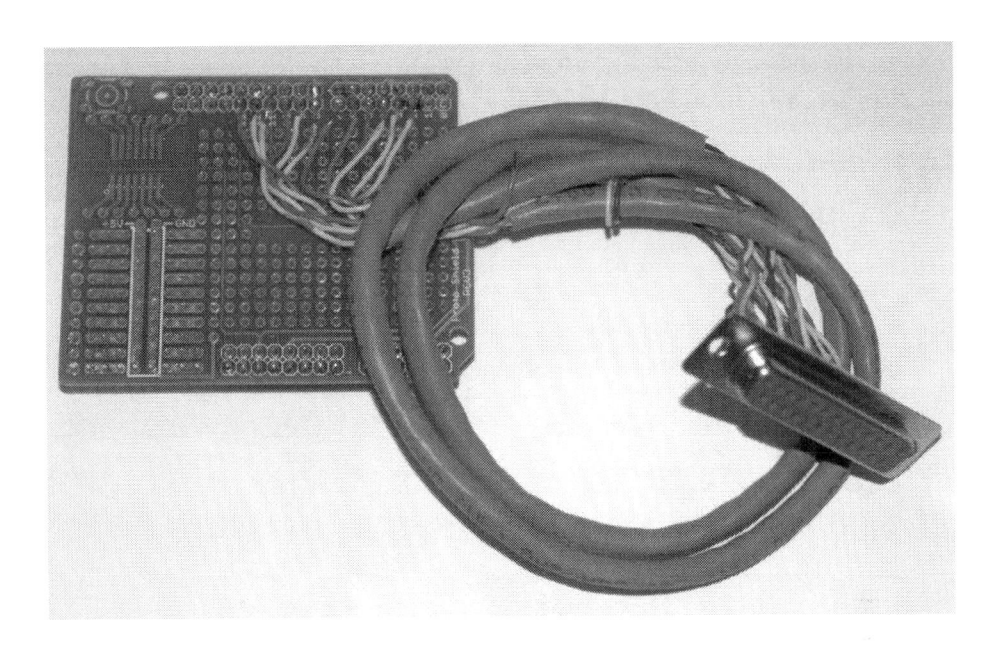

The second option to convert a CNC machine is to integrate the Arduino into the electronics of the CNC machine. To do this you will need access to the "step" and "direction" signals going to each of the stepper motor controllers. The best way to do this is to make a shield with screw terminals that are connected to D2 to D13 and Ground as can be seen in the next picture.

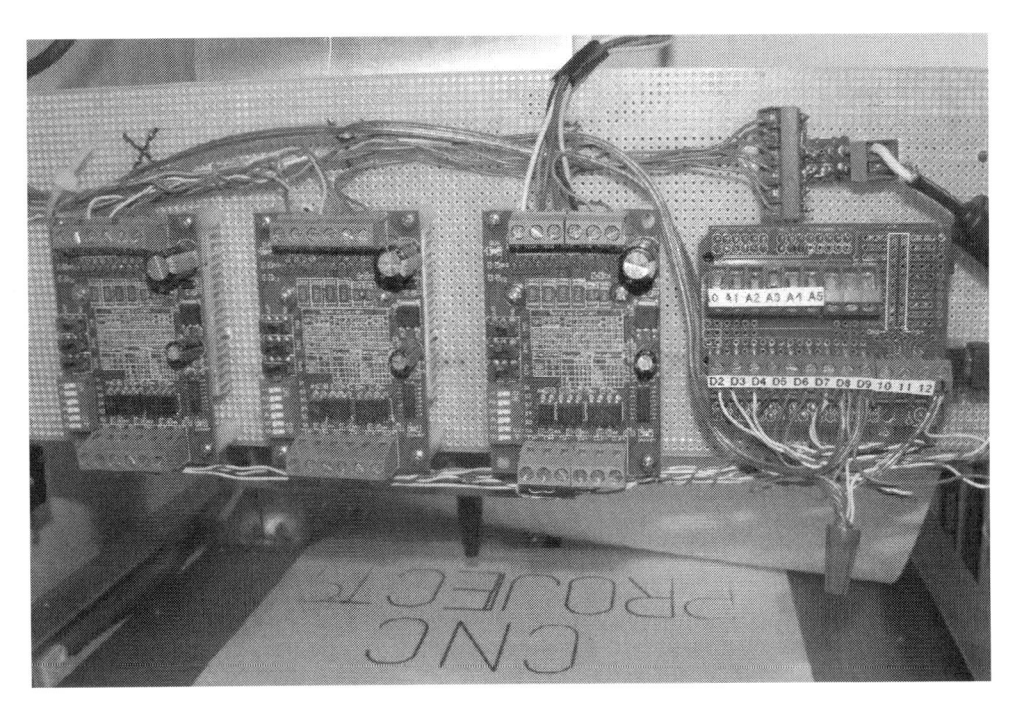

The connections to the Arduino running GRBL are shown in the next picture. Basically, you will need to bring D2 to D12 out to the screw terminals or some other way to easily connect the wires to them. You can also solder the wires to the shield pins like what was done with the 25 pin connector adapter that was shown earlier.

- Ground
- Spindle Dir.
- Spindle Enable
- Z Limit
- Y Limit
- X Limit
- Stepper Enable
- Z Direction
- Y Direction
- X Direction
- Z Step
- Y Step
- X Step

The next schematic shows how to wire the Arduino to the TB6560 driver boards. This design gives full optical isolation between the Arduino and the stepper motor controllers. The X, Y, and Z limit switches are not shown but they are N.O. (Normally Open) and connect to their respective terminals and to ground.

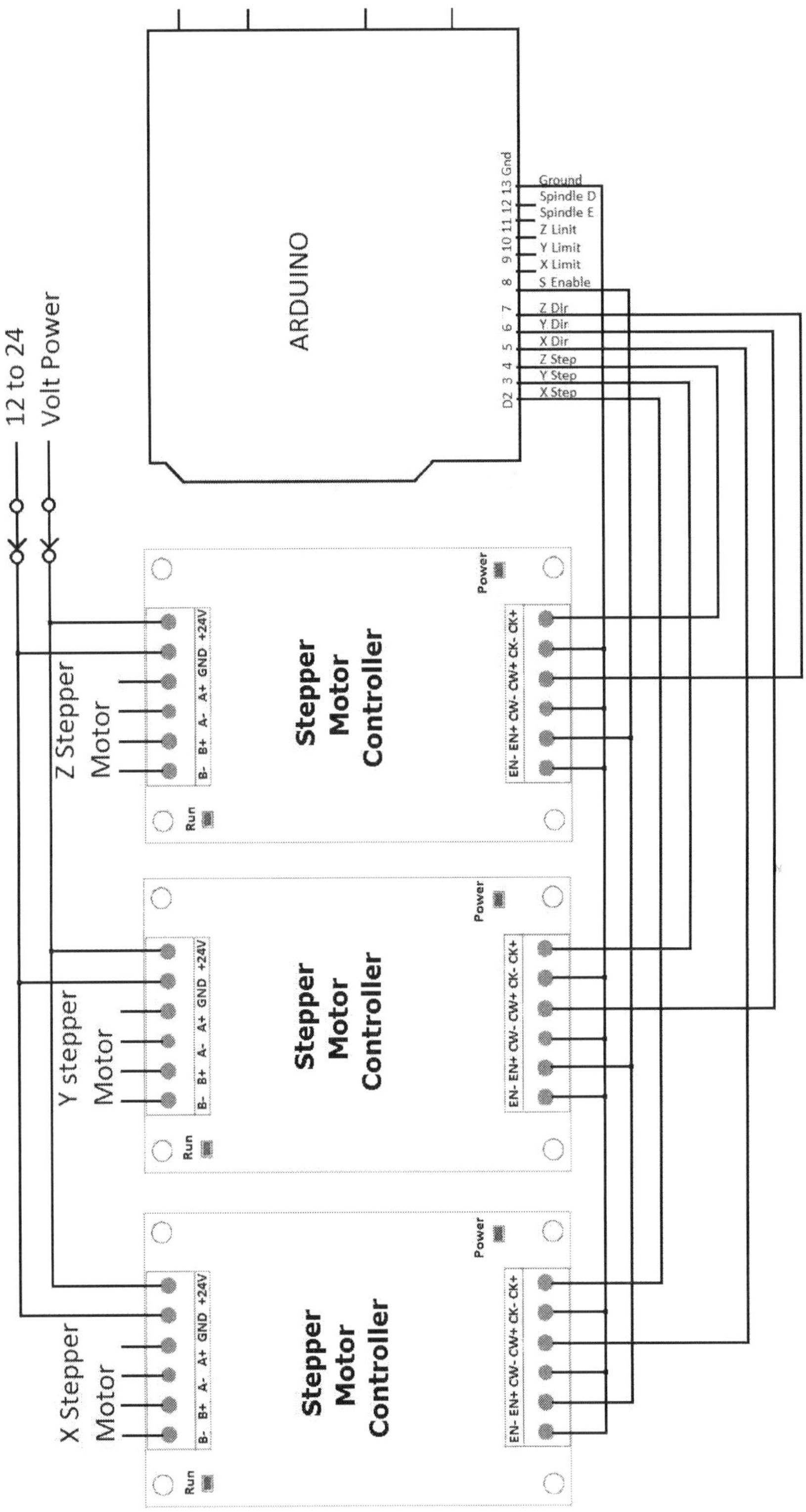
12 to 24
Volt Power
ARDUINO
D2 3 4 5 6 7 8 9 10 11 12 13 Gnd
Ground
Spindle D
Spindle E
Z Linit
Y Limit
X Limit
S Enable
Z Dir
Y Dir
X Dir
Z Step
Y Step
X Step
Z Stepper
Motor
Y stepper
Motor
X Stepper
Motor
Stepper
Motor
Controller
Stepper
Motor
Controller
Stepper
Motor
Controller
B- B+ A- A+ GND +24V
B- B+ A- A+ GND +24V
B- B+ A- A+ GND +24V
EN- EN+ CW- CW+ CK- CK+
EN- EN+ CW- CW+ CK- CK+
EN- EN+ CW- CW+ CK- CK+
Power
Power
Power
Run
Run
Run

The third option is to use a CNC shield like the one in the next picture. The biggest problem with using the CNC shield used to be the lack of sufficient drive power. However, the newer versions that use the DRV8825 IC can deliver up to 2.5 amps. That should be more than enough for smaller CNC machines.

The second problem with the CNC shield is that the outputs are header posts so you might need to make some adapters to screw terminals to be able to connect your stepper motors wires to the CNC shield.

The IC's on the CNC shield run very hot so be sure to attach heat sinks to them. Make sure the heat sinks do not short out any of the pins on the motor controller's connectors.

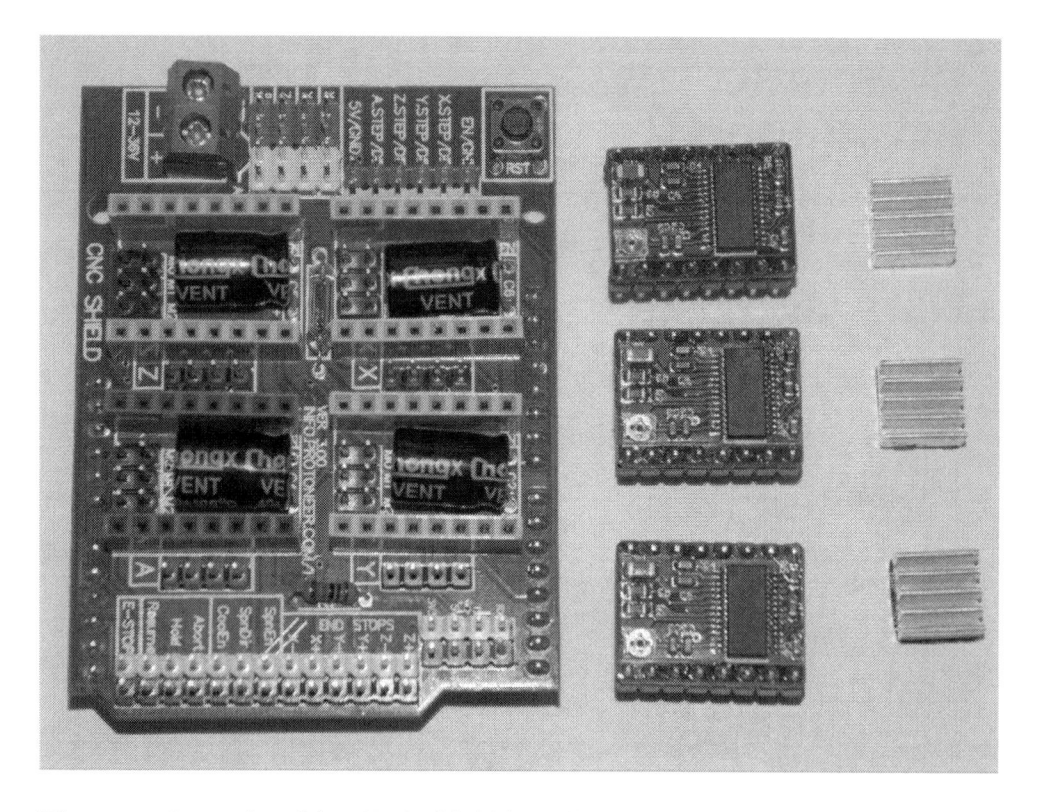

The next picture is of the CNC shield installed on the Arduino. In operation the CNC shield frequently overheated and would shut down the axis until it had cooled off. This made it unusable. I would only recommend it with the smaller NEMA 17 stepper motors like what are used in the 3D printers and laser engravers.

Once you have GRBL loaded and the hardware interfaced you will need to change the default settings to match your CNC machine. The most important settings are to set the inversion of the control signals if needed (S2 and S3 set to 7). When using the external 25 pin adapter method you will likely have to invert them. Another setting is to turn on the limit switches (S21 set to 1). Also, you might want to convert from the default mm to inches (S13 set to 1).

Another critical setting is the steps per mm (or inches). I tried using inches at first but either it created issues or my math was off somewhere. I would recommend using mm instead. For this setting you take the leadscrews number of threads per mm and multiply them by 200 for steps per revolution of the stepper motor. Remember that your leadscrew might have double threads and your stepper motor and stepper controller might be set to 400, 800 or even 1600 steps per revolution. In my case "40" worked as in .2x200 for S100, S101 and S102.

You might have to use trial and error to find the correct settings. Basically, you want the machine to move 1 inch when told to move 25.4 mm. That can be tested with the G code sender program. Set the step size to 25.4 and select "mm". Then click X or Y + or – to see if it moves one inch. If you change the GRBL settings, you usually have to reset the controller to see if the new settings worked. This next image shows my GRBL settings.

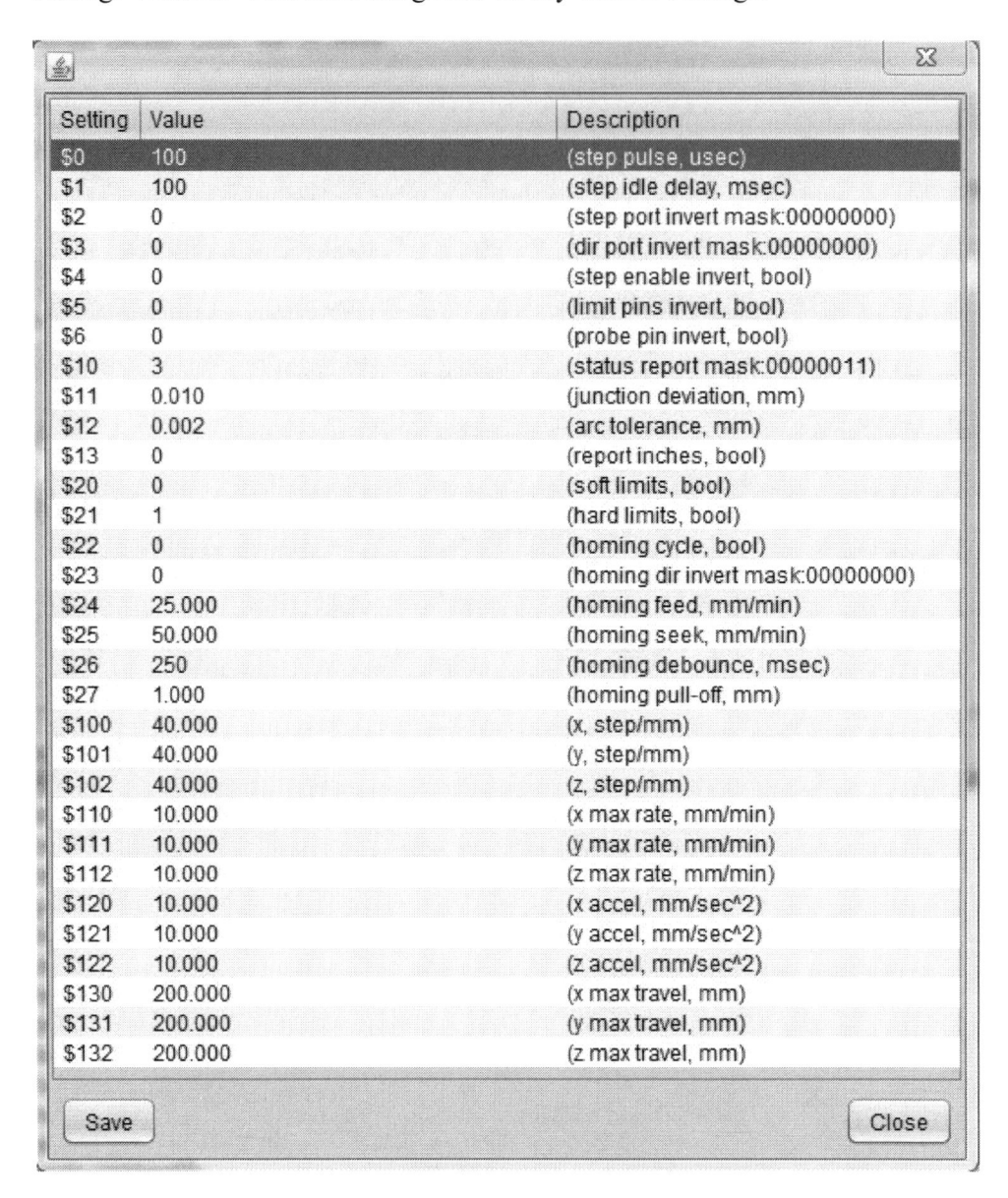

Setting	Value	Description
$0	100	(step pulse, usec)
$1	100	(step idle delay, msec)
$2	0	(step port invert mask:00000000)
$3	0	(dir port invert mask:00000000)
$4	0	(step enable invert, bool)
$5	0	(limit pins invert, bool)
$6	0	(probe pin invert, bool)
$10	3	(status report mask:00000011)
$11	0.010	(junction deviation, mm)
$12	0.002	(arc tolerance, mm)
$13	0	(report inches, bool)
$20	0	(soft limits, bool)
$21	1	(hard limits, bool)
$22	0	(homing cycle, bool)
$23	0	(homing dir invert mask:00000000)
$24	25.000	(homing feed, mm/min)
$25	50.000	(homing seek, mm/min)
$26	250	(homing debounce, msec)
$27	1.000	(homing pull-off, mm)
$100	40.000	(x, step/mm)
$101	40.000	(y, step/mm)
$102	40.000	(z, step/mm)
$110	10.000	(x max rate, mm/min)
$111	10.000	(y max rate, mm/min)
$112	10.000	(z max rate, mm/min)
$120	10.000	(x accel, mm/sec^2)
$121	10.000	(y accel, mm/sec^2)
$122	10.000	(z accel, mm/sec^2)
$130	200.000	(x max travel, mm)
$131	200.000	(y max travel, mm)
$132	200.000	(z max travel, mm)

Chapter 12

Laser Engraving Option

Another option for your CNC machine is to replace the router with a Laser. The laser head is very small in size when compared to the router. The laser that I used required a well regulated 12 volt power source. Because raising and lowering the Laser does not keep it from burning, while the CNC is moving to a new location, you will need to add a circuit to turn the laser on and off. This circuit can also be used to turn off the router when you are done cutting. Basically the circuit consists of a 4N35 optocoupler, a TIP120 or 121 driver transistor and a 12 volt relay. You can easily build this yourself and two of the relay circuits can fit on one small circuit board.

This is the schematic of the relay control board.

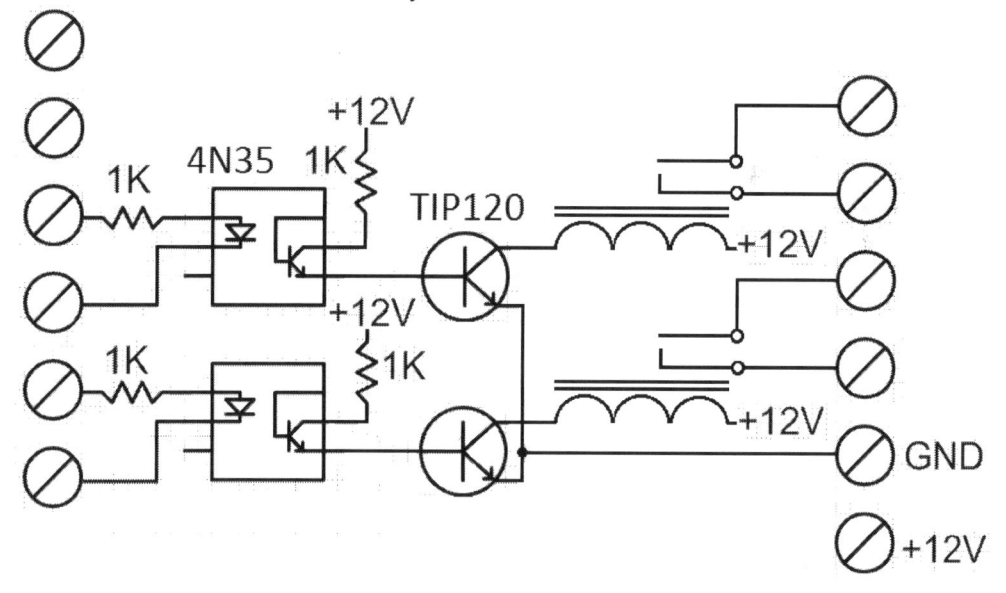

The inputs to the relay board come from D12 and D13 of the Arduino. In my testing with the CNC "M" commands I found that D13 is turned off with a "M03" command and turned on with a "M04" command. For some reason none of the commands toggled the D12 line. So to be able to use your CNC program with a laser you will need to replace the "Z up" commands with a "M03" (laser off) and the "Z down" commands with a "M04" (laser on). You can use Wordpad or Notepad with their "find and replace" function to make the changes automatically. In F-Engrave you can set the Z up to 1 inch and the cut (Z down) depth to 0 inches. Using those settings will make the "find and replace" task easier to do as can be seen in this picture.

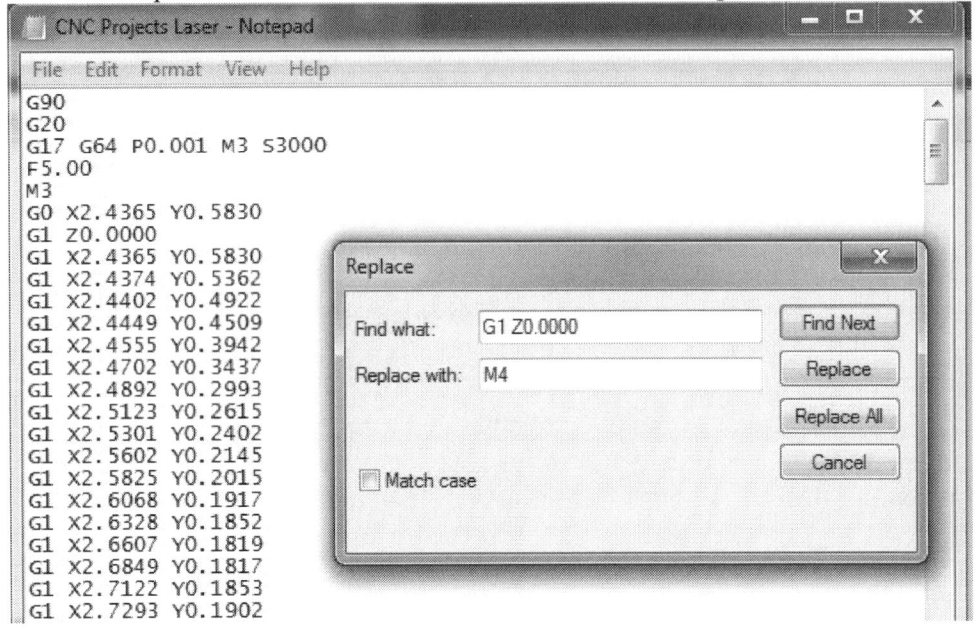

This next picture shows the relay board that is used to control the laser installed to the left of the X axis driver board.

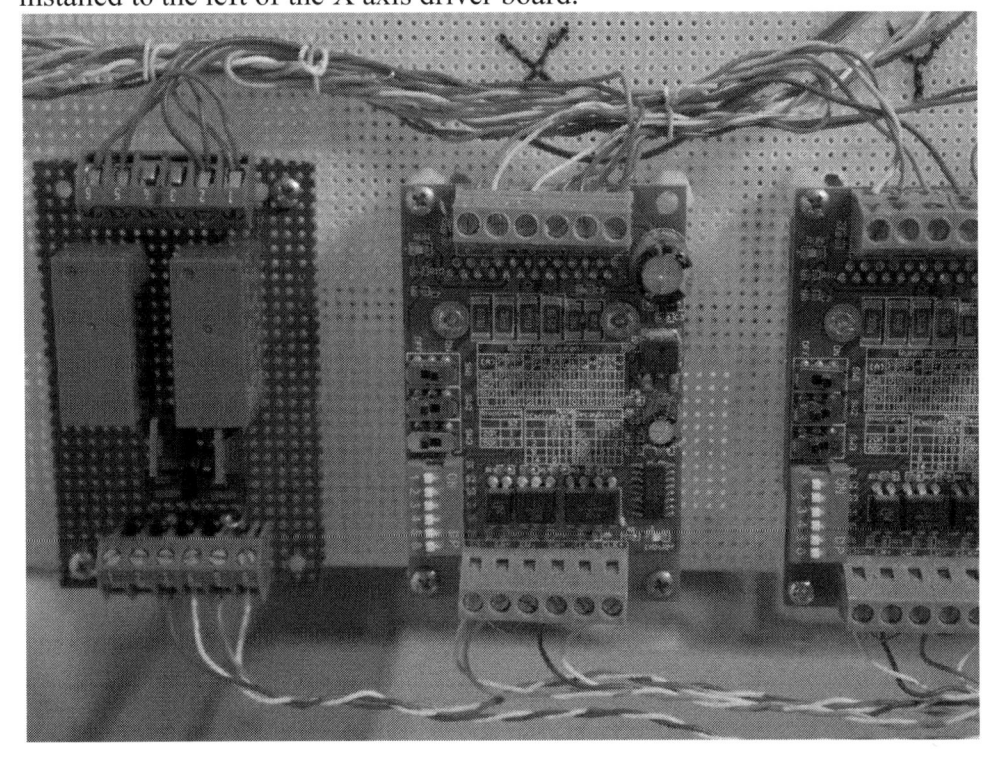

This final picture shows the CNC with the laser working on burning a sign saying "Arduino Books" into some wood.

Bibliography

Computer Numerical Control Programming Basics
A Primer for the Skills USA/VICA Championships
By Steve Krar and Arthur Gill
Distributed by Industrial Press Inc.

LinuxCNC Getting Started V2.3 2014-03-16
Copyright 2000-2012 LinuxCNC.org
By The Linux CNC team
You need to print and read the first 66 pages.

LinuxCNC User Manual V2.5 2014-03-16
Copyright 2000-2012 LinuxCNC.org
By The Linux CNC team
You need to print and read the first 39 pages.
There is lots of CNC Code information in this book.

CNC Web Sites

Here is a guide to many of the available CNC Machine kits.
http://blog.ponoko.com/2011/07/15/pricing-guide-to-diy-cnc-mill-and-router-kits/

DIY CNC Moving Gantry Router at the instructables web site.
http://www.instructables.com/id/DIY-CNC-Router/

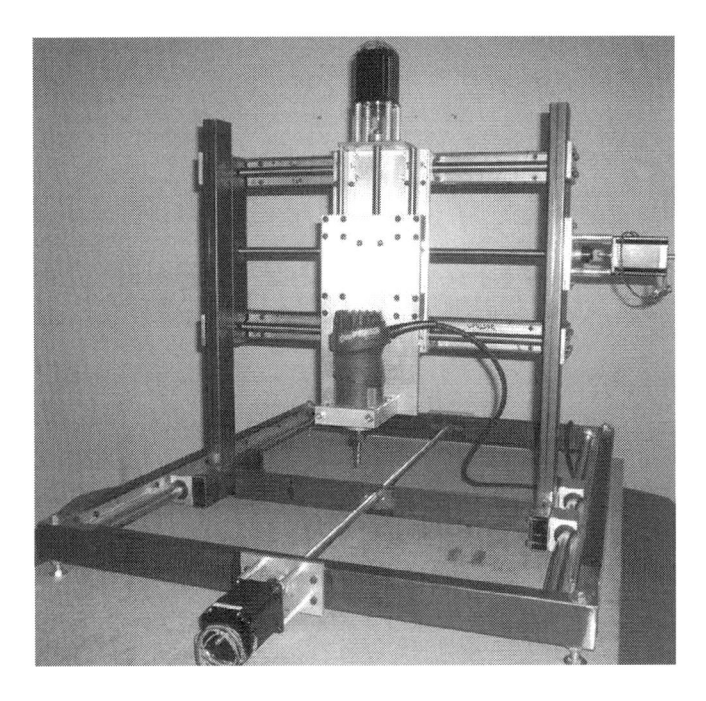

The Zen Toolworks CNC Machine uses a moving Y axis platform. I modeled my machine somewhat after their design.
http://www.zentoolworks.com/

DIY CNC uses a moving gantry.
https://www.kickstarter.com/projects/421256045/diy-desktop-cnc-machine

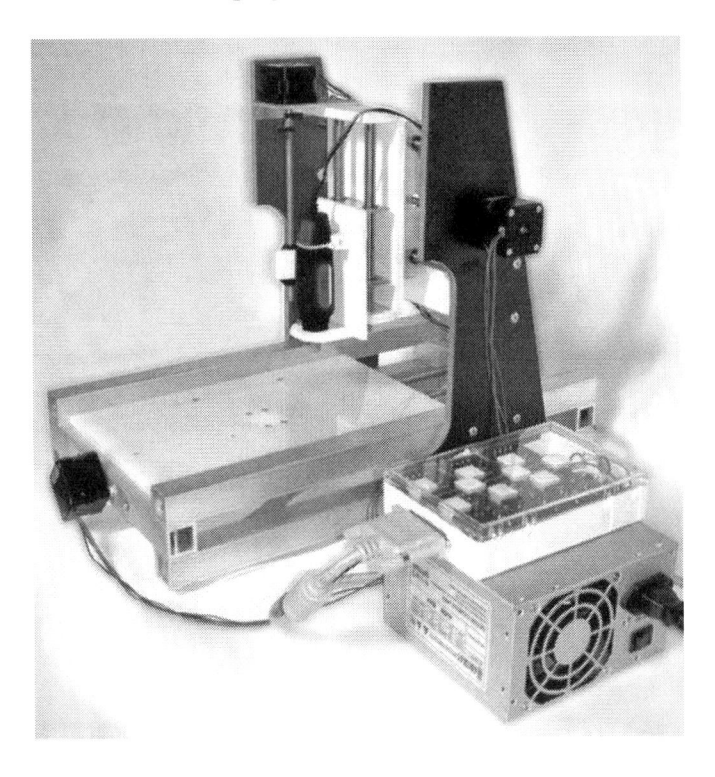

The Fireball V90 assembly instructions on Hacked Gadgets
http://hackedgadgets.com/2009/04/26/fireball-v90-cnc-router-assembly/

This site offers plans for many different CNC machines
https://www.buildyourcnc.com/

A Plexiglas CNC Machine: http://rainbowlazer.com/3d/rhino/diy-3-axis-cnc-millenhanced-machine-controller/

Made in the USA
Middletown, DE
04 March 2024